# XXII National Textile Conference 2020: Traditions and Innovations in Textile and Garment (bgNTC 2020)



IOP Conference Series: Materials Science and Engineering
Volume 1188

Sofia, Bulgaria
12 – 14 November 2020

ISBN: 978-1-7138-4648-2
ISSN: 1757-8981

**Printed from e-media with permission by:**

Curran Associates, Inc.
57 Morehouse Lane
Red Hook, NY 12571

CURRAN ASSOCIATES INC. proceedings.com

**Some format issues inherent in the e-media version may also appear in this print version.**



Institute of Physics
Dirac House, Temple Back
Bristol BS1 6BE UK

Phone: 44 1 17 929 7481
Fax: 44 1 17 920 0979

techtracking@iop.org

**Additional copies of this publication are available from:**

Curran Associates, Inc.
57 Morehouse Lane
Red Hook, NY 12571 USA
Phone: 845-758-0400
Fax: 845-758-2633
Email: curran@proceedings.com
Web: www.proceedings.com

# TABLE OF CONTENTS

IOP Conf. Series: Materials Science and Engineering 1188 (2021) 011001 doi:10.1088/1757-899X/1188/1/011001

SEUTGL Preface

**XXII National Textile Conference (XXIIbgNTC2020), 12-14.11.2020 - Sofia**

The National Textile Conference on "Traditions and Innovation in Textiles and Garment" have been held for the 22nd time from 12 to 14 November 2020 in the National house of science and engineering – Sofia.

Main organizer of the conference is Scientific Engineering Union of Textiles, Garment and Leathers (**SEUTGL**). The main partner of **XXIIbgNTC2020** is the Federation of Scientific Engineering Unions in Bulgaria.

Within the framework of the research association, the co-organizers of the conference were the Technical University - Sofia, the Chemical Technology and Metallurgical University - Sofia, as well as the Institute of Polymers at the Bulgarian Academy of Sciences. Partners of the conference events are the educational units of the higher and secondary schools in the country, as well as industrial enterprises.

The Scientific Committee of **XXIIbgNTC2020** with Chair - Professor Diana Germanova-Krasteva and the main organizer SEUTGL with Chair – Associate Professor Ivelin Rahnev provided excellent conditions for the presentation of the various works and discussions in the sidelines. Over 30 university lecturers, researchers, experts, PhD students and students are participating in the scientific sessions of the forum.

Out of the 25 abstracts submitted in the program of the Scientific Sessions, 15 reports and 10 posters on the current trends in the technology of textiles, clothing and design were presented.

All reported works reflect the efforts of research, university and industrial teams from the specialized departments and laboratories of TU - Sofia, UCTM - Sofia, NBU - Sofia, National Academy of Arts - Sofia, Institute of Polymers of the Bulgarian Academy of Sciences, Defense Institute - Sofia, SWU - Blagoevgrad, VFU - Varna and MBBU - Botevgrad, PGTO Dobri Zhelyazkov - Sliven, NHG Dimitar Dobrovich - Sliven, and the Museum of Textile Industry in Sliven.

In the Science Committee participated leading lecturers from the University of Southern Alsace in Mulhouse, France, Dresden Technical University, Germany, Hanoi University of Science and Technology, of Vietnam, University "Ss. Cyril and Methodius"- Skopje, Republic of North Macedonia and the National School of Engineering of Monastir, Tunisia.

The main support for the good organization and technical implementation of the planned activities at XXIIbgNTC2020 comes from E. Miroglio EAD - Sliven, Picanol nv from Belgium, Pirin Tex EOOD - Gotse Delchev and Lempriere Wool EOOD - Sliven.

The reports submitted at **XXIIbgNTC2020** will be published sequentially in the issues of the Textiles and Garment magazine *ISSN1310-912X* (print), *ISSN 2603-302X* (online), which will

Sliven, 14.11.2020

IOP Conf. Series: Materials Science and Engineering 1188 (2021) 011001 doi:10.1088/1757-899X/1188/1/011001



be available to readers in university and public libraries as well as to the websites of SEUTGL: www.tok.fnts.bg and the publishing house of the association: www.bgtextilepublisher.org.
Selected papers were presented to the publisher IOP Conference Series: Materials Science and Engineering in order to be indexed in World of Science data base.
Sofia, November 14th 2020

Compiled by:
Associate Professor Ivelin Rahnev, Chair of SEUTGL

IOP Conf. Series: Materials Science and Engineering 1188 (2021) 011001 doi:10.1088/1757-899X/1188/1/011001

SEUTGL Preface

# SCIENTIFIC COMMITTEE

Of the 22nd National Textile Conference with international participation, Sofia, Bulgaria
November 12th to 14th 2020

## HONORARY MEMBERS

1. Prof. Ivan GEORGIEV, PhD, Technical University – Sofia
2. Prof. Iliya RASHKOV, DSc, corr.-member of the Bulgarian Academy of Sciences, IP – Sofia
3. Prof. Nikolay SIMEONOV, DSc, University of Chemical Technology and Metallurgy – Sofia
4. Prof. Hristo PETROV, PhD, Technical University – Sofia
5. Prof. Penka Dimitrova, PhD, Technical University – Sofia
6. Prof. Margarita NEZNAKOMOVA, PhD, Technical University – Sofia
7. Prof. Vihroni POPNEDELEV, National Academy of Arts – Sofia
8. Assoc. Prof. Anna GEORGIEVA, PhD, University of Chemical Technology and Metallurgy – Sofia
9. Assoc. Prof. Rumen RUSSEV, PhD, Trakia University – Stara Zagora

## FOREIGN MEMBERS

10. Prof. Jean–Yves DREAN, DSc, Université de Haute-Alsace -ENSISA-LPMT, Mulhouse, France
11. Prof. Alenka Majcen Le MARESHAL, DSc, University of Maribor, Slovenia
12. Prof. Dr. Yordan KYOSEV, Prof. Dr. Ing. habil, TU – Dresden, Germany
13. Prof. Goran DEMBOSKI, PhD, University "Ss. Cyril and Methodius" – Skopje, Macedonia
14. Assoc. Prof. VU Thi Hong Khanh, PhD, Hanoi University of Science and Technology – STLF, Vietnam
15. Prof. Saber Ben ABDESSALEM, PhD, Ecole Nationale d'Ingeneurs de Monastir, Monastir, Tunisie
16. Dr Gaetano RIMINI, "E. Miroglio" EAD – Sliven

## RESPONSIBLE REVIEWERS OF THE SCIENTIFIC PROGRAM

17. Prof. Diana GERMANOVA-KRASTEVA, PhD, TU – Sofia, Chair of the Conference
18. Prof. Maya BOGDANOVA, National Academy of Arts (NAA) – Sofia
19. Prof. Andreas CHARALAMBUS, PhD, TU – Sofia
20. Prof. Snezhina ANDONOVA, PhD, South-West University – Blagoevgrad
21. Prof. Adelina Popnedeleva, PhD, National Academy of Arts – Sofia
22. Prof. Radostina ANGELOVA, DSc, TU – Sofia
23. Assoc. Prof. Maria SPASOVA, PhD, IP - BSA – Sofia
24. Assoc. Prof. Kapka MANASSIEVA, PhD, VFU – Varna
25. Assoc. Prof. Stela BALTOVA, PhD, International Business School – Botevgrad
26. Assoc. Prof. Zlatina KAZLACHEVA, PhD, Trakia University - Stara Zagora
27. Assoc. Prof. Desislava Staneva GRABCHEVA, PhD, UCTM – Sofia
28. Assoc. Prof. Krassimir DRUMEV, PhD, TU – Gabrovo
29. Assoc. Prof. Ivelin RAHNEV, PhD, E. Miroglio EAD – Sliven.

Sofia, June 2020

On behalf of the scientific engineering union of Textile, Garment and Leathers:

Assoc. Prof. Ivelin RAHNEV, PhD

Sliven, 14.11.2020

# Peer Review Declaration Form Guidance

On the following page, you will find the declaration form.

- Please answer each question.
- You should submit the form along with the rest of your submission files.
- The deadline is the submission date written in your publishing agreement.

**All conference organisers/editors are required to declare details about their peer review.**

**We will published the information you provide as part of your proceedings.**

IOP Conf. Series: Materials Science and Engineering 1188 (2021) 011002 doi:10.1088/1757-899X/1188/1/011002

# Peer review declaration

All papers published in this volume of IOP Conference Series: Materials Science and Engineering have been peer reviewed through processes administered by the Editors. Reviews were conducted by expert referees to the professional and scientific standards expected of a proceedings journal published by IOP Publishing.

- **Type of peer review: Single-blind / Double-blind / Triple-blind / Open / Other (please describe):**
  **Triple-blind**
- **Conference submission management system:**
  **Announcement of the conference on the website of the main organizer, submission of materials by e-mail.**
  **Distribution of articles for review by e-mail.**
- **Number of submissions received:**
  **35**
- **Number of submissions sent for review:**
  **35**
- **Number of submissions accepted:**
  **8**
- **Acceptance Rate (Number of Submissions Accepted / Number of Submissions Received X 100):**
  **23**
- **Average number of reviews per paper:**
  **3**
- **Total number of reviewers involved:**
  **20**
- **Any additional info on review process:**
  **Communication by emails**
- **Contact person for queries:**
  Name : Ivelin Rahnev
  Affiliation: Scientific Engineering Union of Textiles, Garment and Leathers
  Email : tok.chair@fnts.bg

*Sliven, September 22nd 2021*
*Rahnev*

# Synthesis and characterisation of a new water soluble fluorescent cationic polymer and its microbiological activity

**Hristo Manov[1], Desislava Staneva[2], Evgenia Vasileva-Tonkova[3], Ivo Grabchev*[4]**

[1]Sofia University "St. Kliment Ohridski", Faculty of Chemistry and Pharmacy, 1407 Sofia, Bulgaria

[2]University of Chemical Technology and Metallurgy, 1756 Sofia, Bulgaria

[3]The Stephan Angeloff Institute of Microbiology, Bulgarian Academy of Sciences, 1113 Sofia, Bulgaria

[4]Sofia University "St. Kliment Ohridski", Faculty of Medicine, 1407 Sofia, Bulgaria

i.grabchev@chem.uni-sofia.bg

**Abstract**. The work describes the synthesis of a new fluorescent cationic polymer (PNI) containing 1,8-naphthalimide units. Its photophysical characteristics have studied in organic solvents and water. The investigations have revealed, the polymer in solution to have an intense yellow color and to emit yellow-green fluorescence. These properties have been preserved after depositing it onto a cotton fabric. The color coordinates (L*, a*, b* and XYZ) of the dyed cotton fabric have been determined. The antimicrobial activity of PNI has been studied in a liquid medium against bacterial strains *Bacillus cereus* and *Pseudomonas aeruginosa*, and the fungal strain *Candida lypolitica*. The results obtained show moderate activity of the polymer against the tested strains which had been preserved after its loading onto the cotton fabric.

## 1. Introduction

Due to the compact molecular structure and low molecular weight, 4-amino substituted 1,8-naphthalimide fluorophores are one of the most studied organic fluorophores with intense yellow-green color for dyeing textile and polymeric materials [1]. In this case, the nuance of the color and fluorescence intensity depends on the polarization of the chromophore system, which is predetermined by the electron-donor capacity of the substituent at the C4 position. Thus, by varying the type of substituents, fluorophores with tailored properties can be obtained [2-6]. On the other hand, by introducing special substituents attached to the imide nitrogen atom and to the C4 atom of the chromophore system, 1,8-naphthalimide derivatives can be used as fluorophores for structural coloring of polymeric materials. Thus they are covalently attached to the polymer chain and no migration from the surface or inside of the polymers is observed [1]. Derivatives of 1,8-naphthalimide can also participate in polycondensation processes with some traditional monomers [1]. Polypropylene imine and polyamidoamine dendrimers modified with 1,8-naphthalimide units acquire interesting photophysical characteristics and can be used in the design of highly efficient sensor systems for detecting metal ions and protons [7-10].

In recent years, the antimicrobial activity of 1,8-naphthalimide derivatives has been intensively studied [11-14]. The incorporation of such compounds into polymer or dendrimer macromolecules leads to an increase in their biological activity [15-17]. This activity is preserved after their deposition onto textile materials, which gives them new valuable properties [18,19]. Such textiles with antibacterial properties are widely used in for domestic purposes, industry, automotive plants in order to reduce the growth of pathogenic bacteria. Their application in clinical practice for the production of bed linen, wound dressings, surgical sutures, etc. has been especially relevant.

The aim of this work is to synthesize a new polymer with quaternary groups and to study its photophysical characteristics in organic solvents. The antimicrobial activity of the polymer and the colored cotton fabric have also been assessed.

## 2. Experimental part

### 2.1. *Materials and methods*

The synthesis and characterization of the initial monomer unit: 2-(pyridin-4-ylmethyl)-6-((pyridin-4-ylmethyl)amino)-1H-benzo[de] isoquinoline-1,3(2H)-dione (NI) was described in [20]. Organic solvents of spectroscopic grade (dioxane; chloroform; acetonitrile; dimethylsulfoxide (DMSO); *N,N*-dimethylformamide (DMF); tetrahydrofurane (THF); ethanol (EtOH) and methanol (MeOH)); and and sebacoyl chloride (99%) were used as obtained from Sigma Aldrich.

The photophysical measurements were performed on a Thermo Spectronic Unicam UV 500 spectrophotometer and on a Cary Eclipse spectrofluorometer. Thin layer chromatography was carried out on silica gel (Fluka F60 254 20x20; 0.2 mm) using an eluent system acetone/n-heptane (1:1). The molecular weight of polymer was determined on a GPC Water 244 apparatus with polystyrene standard and THF as an eluent. Both a differential refractive index and an UV-vis absorption detector ($\lambda_A$ = 430 nm) were used. 1H NMR (600.13 MHz) and 13C (150.92 MHz) spectra were acquired on an AVANCE AV600 II+ NMR spectrometer. Fourier-transform spectrometer (IRAffinity-1 Shimadzu, Kyoto, Japan) at a 2 cm−1 resolution. Color coordinates of the dyed cotton fabrics were determined using Datacolor Spectraflash SF300 spectrophotometer (Datacolor, NJ, USA).

### 2.2. *Synthesis of Polymer (PNI)*

NI (3.94 g, 0.01mol) was dissolved in 100 ml anhydrous acetone and sebacoyl chloride (4.80 ml, 0.02 mol) was added dropwise to the solution at 5°C. The mixture was stirred for 30 minutes. Then the precipitate was filtered and washed with acetone (3 times 50 ml). Yield: 8.50 g
FT-IR (KBr) $cm^{-1}$: 3222, 3072, 3066, 2993, 2944, 1683, 1648, 1633, 1575, 1503, 1356, 1299, 1240, 998, 776, 755. $^{1}$H NMR ($CDCl_3$, δ, ppm): 8.35 (d; J=7.12 Hz; 4H); 8.28 (d; J=8.12 Hz; 1H); 8.20 (d; J=8.05 Hz, 1H); 7,93. (bs; 1H); 7.54 (d, J=8.14 Hz; 4H); 7.20 (d, J=8.32 Hz; 1H); 7.06 (s, 1H); 4.30 (s; 1H); 2.20 (t, J=6.19 Hz; 4H); 2.02 (m, 8H); 140(m, 8H); $^{13}$C NMR ($CDCl_3$, δ, ppm): 176.4, 164.5, 163.9, 154.3, 143.2, 140.3, 137.1, 134.2, 133.0, 130.5, 128.4, 123.4, 118.3, 113.0, 48.0, 42.1, 29.3. 28.1.

### 2.3. *Dyeing of cotton samples with PNI*

1.0 g of cotton fabric (140 g/$m^2$) was immersed into 10 ml aqueous solution of PNI (10 mg) at 25°C for 30 min. After that the sample was dried at room temperature.

### 2.4. *Antimicrobial activity test*

#### 2.4.1. *Minimum Inhibitory Concentration.*

Broth dilution test was used for determination of the minimum inhibitory concentrations (MICs) of PNI against Gram-positive *B. cereus* and Gram-negative *P. aeruginosa* and the yeasts *C. lypolitica*. The polymer was serially diluted in tubes with meat-peptone broth (MPB). The tubes were inoculated with respective microbial suspension and incubated at appropriate temperature for 24 h. The microbial growth was assessed by the turbidity at 600 nm ($OD_{600}$). The growth control, sterility control and sample control

IOP Conf. Series: Materials Science and Engineering 1188 (2021) 012001 doi:10.1088/1757-899X/1188/1/012001

were used. The lowest concentration of the samples that inhibited the visible growth of the strains was referred as MIC. Three independent experiments were carried out and averages are given.

### *2.4.2. Antibacterial activity of the treated cotton fabric.*

The antibacterial activity of cotton fabric treated with PNI was tested against *B. cereus* and *P. aeruginosa* as model strains. Test tubes containing MPB and square cotton speciments (10 mm x 10 mm) were inoculated with cell suspension of each bacterial culture. Tubes with untreated cotton and without speciments were also prepared as controls. After incubation for 24 h, the specimens were removed and $OD_{600}$ was determined. The antimicrobial activity of the treated cotton samples was evaluated by the reduction of $OD_{600}$ after incubation compared to the control sample. All antimicrobial tests were done in triplicate and the average was taken; standard deviations were less than 5%.

## 3. Results and discussion

A new fluorescent polymer with quaternary amonium groups has been obtained by a reaction between NI containing two pyridinium nuclei and a bifunctional sebacoyl chloride. In the reaction between them, the acyl chloride attacks the pyridinium nitrogen atom to form a quaternary ammonium group (Scheme 1). The reaction between the co-monomers was carried out at 5°C for 30 minutes, whereby the polymer PNI was obtained in almost quantitative yield as a bright yellow solid.

The molecular weight of the polymer Mn = 8 940 Da and the polydispersity Mw/Mn = 1.54 were determined. On the basis of the molecular weight, the degree of polymerization DP has been calculated as a ratio of molecular weight of the polymer (Mn) and molecular weight of the repeat units in the polymer structure (Mo) according to the formula DP=Mn/Mo. The results obtained have showed that, approximately eight NI units are linked in the polymer structure.

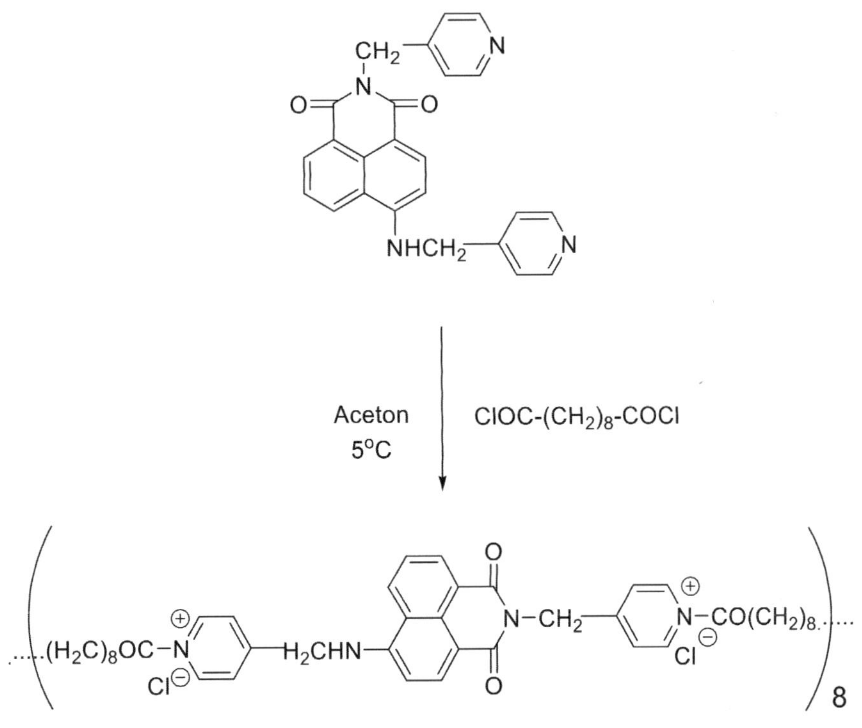


Scheme 1. Synthesis of fluorescent PNI

Figure 1 presents comparatively the IR spectra of the initial NI and the resulting polymer PNI. The characteristic bands for the carbonyl groups of NI are at 1674 $cm^{-1}$ and 1636 $cm^{-1}$, while for PNI they are slightly hypsochromically shifted and are at 1683 $cm^{-1}$ and 1638 $cm^{-1}$, respectively. That can be explained with the presence of a positive charge at the nitrogen atom of the pyridine nuclei. There is a broadening of the signal and a shoulder at 1662 $cm^{-1}$ in the spectrum of PNI, what is due to the oscillation of the N-C = O group as this signal overlaps with the intense peak of C = O. An intense peak has been registered at 1501 $cm^{-1}$, which refers to the oscillation of the $–CH_2–$ groups from the polymer structure. As seen from Figure 1, this peak is missing in the spectrum of NI.

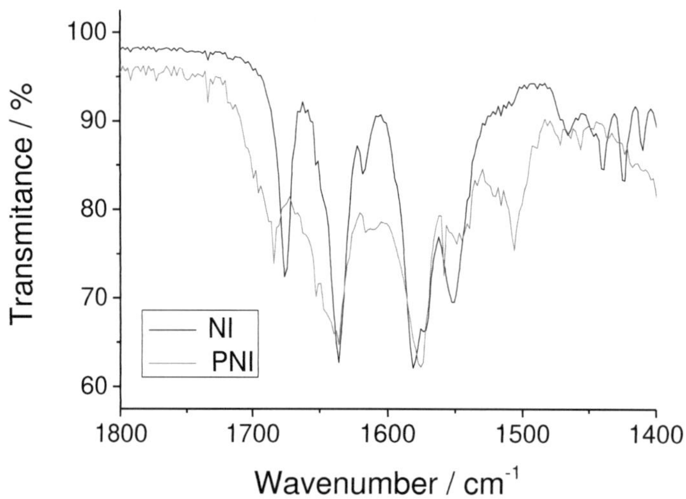


Figure 1. Infrared spectra of NI and PNI

Table 1. Photophysical characteristics of PNI in different solvents

| Solvents | $\lambda_A$ nm | $\lambda_F$ nm | $\nu_A$-$\nu_F$ $cm^{-1}$ | $\Phi_F$ |
|---|---|---|---|---|
| DMF | 430 | 515 | 3934 | 0.38 |
| Water | 437 | 533 | 4122 | 0.15 |
| Ethanol | 430 | 519 | 3988 | 0.36 |
| Dioxane | 414 | 498 | 3505 | 0.26 |
| Methanol | 428 | 523 | 4244 | 0.29 |
| Chloroform | 424 | 502 | 5142 | 0.25 |
| DMSO | 425 | 511 | 3868 | 0.22 |
| THF | 417 | 502 | 4060 | 0.26 |
| Acetonitrile | 415 | 508 | 4411 | 0.24 |

The main photophysical characteristics of PNI have been studied in organic solvents of different nature and polarity. The presence of quaternary groups and positive charges in the structure of the PNI allows this polymer to dissolve well in water. In the solvents used PNI has an intense yellow color and the emitted fluorescence is yellow-green. The data in Table 1 show that, the absorption maxima are in the 415-437 nm range and the corresponding fluorescent maxima are at 498-533 nm. Figure 2A shows the dependence of these maxima on the polarity of the medium determined using the parameter $E_T(30)$. The maxima are batochromically shifted in both cases comparing those in the polar environment to the ones in non polar, which means that PNI exhibits positive solvatochromism.

The fluorescence quantum yield ($\Phi_F$) has been calculated on the basis of the absorption and fluorescence spectra by equation 1.

$$\Phi_F = \Phi_{st} \frac{S_u}{S_{st}} \frac{A_{st}}{A_u} \frac{n_{Du}^2}{n_{Dst}^2} \quad (1),$$

where the $\Phi_F$ is the emission quantum yield of the sample; $\Phi_{st}$ is the quantum yield of the standard Rhodamine ($\Phi_{ref}$= 0.94 [21]); $A_{st}$ and $A_u$ are the absorbance of the standard and sample at the excited wavelength, respectively; while $S_{st}$ and $S_u$ are the integrated emission band areas of the standard and sample, at the excited wavelength respectively, and $n_{st}$ and $n_u$ are the solvent refractive indexes of the standard and sample; subscripts *u* and *s* refer to the unknown and standard, respectively.

As Figure 2B shows, the fluorescence intensity also depends on the polarity of the medium and on its ability to form hydrogen bonds with the NI chromophore system. Higher intensity has been registered in the polar solvents, where the dipole-dipole interactions and the formed hydrogen bonds lead to a change in the polarization of the NI. The calculated quantum yield of fluorescence shows that, the polarity of the medium affects also its values. In organic solvents $\Phi_F$ is in the 0.22-0.38 region, while in aqueous solution it is 0.15.

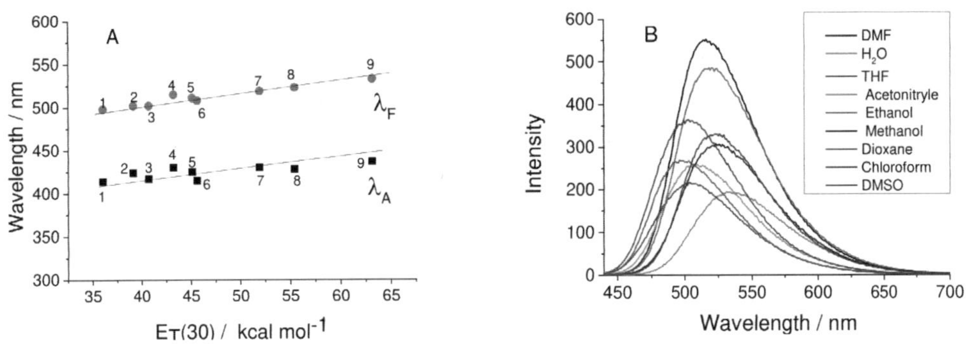


Figure 2. (A) Dependence of absorption and fluorescence maxima of PNI on the empirical parameter of solvent polarity ET(30): 1-dioxane; 2-chloroform; 3-THF; 4-acetonitrile; 5-DMSO; 6-DMF; 7-Ethanol; 8-Methanol; 9-water and (B) fluorescence spectra of NI in the same solvents.

The deposition of PNI on a cotton surface has been carried out after soaking the fabric with 1% aqueous polymer solution at room temperature for 30 minutes. The cotton fabric dyed with PNI is yellow in color and emits yellow-green fluorescence. The degree of loading has been determine by fluorescence spectroscopy and it has been found that, 89% of PNI used had been deposited onto the cotton surface. When PNI is deposited on the cotton fabric, the polymer is retained by the formation of hydrogen bonds and other weak non-covalent interactions between the fabrics and polymer. In this case, the hydrogen bonds have been formed with the unshared electron pairs of the carbonyl groups of the polymer and the hydroxyl groups of the cellulose structure of the cotton materials.

The color coordinates have been determined using Datacolor system: CIEab coordinates (L*, a* and b*) and chromaticity coordinates (XYZ, and xy), which have been compared with those of the untreated cotton fabric. The data obtained are summarized in Table 2. The results show that, the cotton fabric used is white, and acquires a saturated and bright yellow color after been traded with PNI.

Table 2. Color characteristics of the non-treated and treated with PNI cotton fabric

| | L* | a* | b* | X | Y | Z | x | y |
|---|---|---|---|---|---|---|---|---|

IOP Conf. Series: Materials Science and Engineering 1188 (2021) 012001 doi:10.1088/1757-899X/1188/1/012001

| | | | | | | | | |
|---|---|---|---|---|---|---|---|---|
| Non treated cotton | 93.77 | -0.26 | 3.75 | 80.22 | 84.75 | 85.65 | 0.3201 | 0.3382 |
| Cotton–PNI | 77.64 | 6.73 | 57.94 | 52.42 | 52.61 | 14.88 | 0.4372 | 0.4388 |

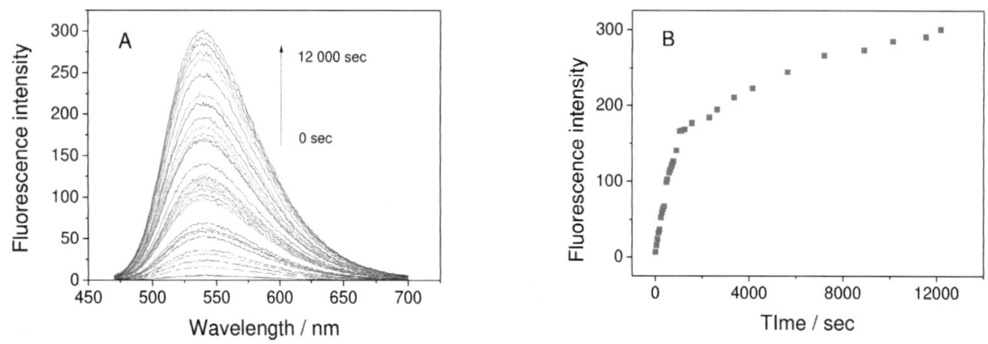


Figure 3. Fluorescence spectra (A) and change in the fluorescence intensity (B) over time after immersion of cotton–PNI into an aqueous solution

When the cotton fabric is immersed into an aqueous solution the polymer is slowly released from the surface of the cotton matrix, due to the very good water solubility of PNI. The release process has been registered by fluorescence spectroscopy. Figure 3 shows the fluorescence spectra tracking the release of PNI with the time of up to 12,000 sec. (Figure 3A) and the dependence of their intensity on time (Figure 3B). Up to 1200 seconds the release is more intense, then up to 12,000 sec., it slows down.

The antimicrobial activity of PNI has been tested against Gram-positive bacteria, Gram-negative bacteria and fungal *Candida* strain. The method of dilution in the concentration range $2x10^{-6}$ -$1x10^{-3}$M) in MPB has been used to determine MICs of PNI. It has been found that, the inhibition of the growth of Gram-positive bacteria *B. cereus* is higher (MIC = 33.3 μM) than that of Gram-negative bacteria *P. aeruginosa* (MIC-44.7 μM), while in fungal strain C. *lipolytica* the activity was lower (MIC = 56.0 μM). The difference in the activity of PNI against the tested bacteria can be explained with the different structure of Gram-positive and Gram-negative bacteria.

The antibacterial activity of cotton fabrics treated with PNI (10/10 mm) against *P. aeruginosa* and *B. cereus* has been tested after immersion of the cotton sample into MPB. Figure 4 shows that, the reduction in bacterial growth is better expressed in the case of Gram-positive *B. cereus*, where the reduction is about 26%, while in Gram-negative bacteria it is 11%. In this case, the effect is due to the possibility of PNI to release from the fabric, what facilitates its direct contact with bacterial cells.

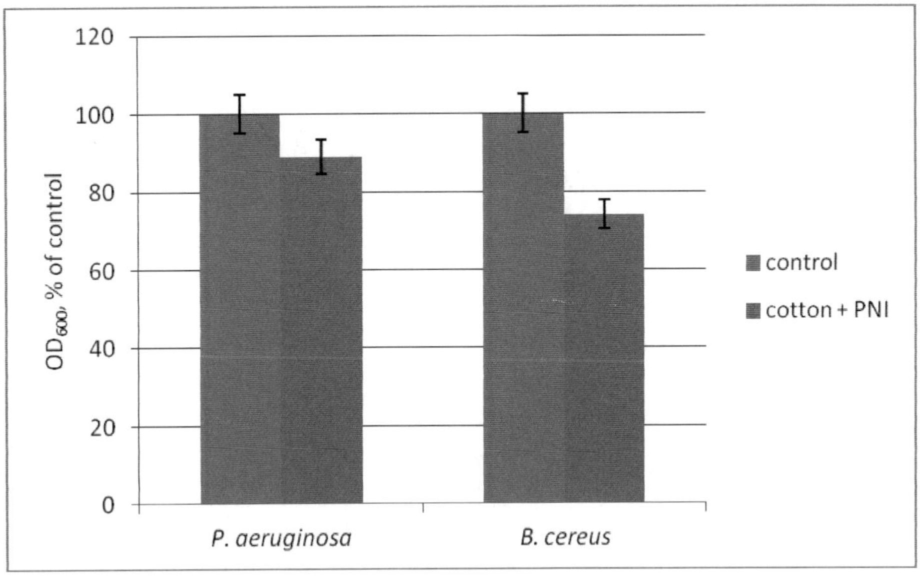


Figure 4. Reduction of the growth of model bacteria by treated with PNI cotton fabrics.

## 4. Conclusion

A new water-soluble cationic polymer, which emits intense yellow-green fluorescence has been obtained and characterized. The determined photophysical characteristics in organic solvents and water have demonstrated the absorption maxima to be in the 415-437 nm range and the corresponding fluorescent maxima in the 498-533 nm range with a well-defined positive solvatochromism. The quantum yield in organic solvents is in the range 0.22-0.38, while in aqueous solution it is lower. The microbiological activity of the polymer in solution and after its application onto a cotton fabric has been studied. It has been found that, the inhibition of the growth of Gram-positive bacteria *B. cereus* (MIC = 33.3 μM) is higher than that of Gram-negative *P. aeruginosa* (MIC-44.7 μM), while in the case of yeasts *C. lipolytica* the activity is lower (MIC = 56.0 μM).

## Acknowledgements

This work was supported by Grants КП-06-КОСТ/19 from the National Science Fund, Ministry of Education and Science of Bulgaria.

## References

[1] Dodangeh M, Grabchev I, Staneva D and Gharanjig K 2021 *Fibers. Polym* (in press)
[2] Grabchev I, 1998 *Dyes Pigm.* **38** 219-226.
[3] Dalius Gudeika, 2020 *Synthet. Met.* **262** 116328
[4] Magalhaes J L, Pereira RV, Triboni ER, Berci Filho P, Gehlen M H and Nart F C 2006 *J. Photochem. Photobiol. A: Chem.* **183** 165-170
[5] Zagranyarski Y, Mutovska M, Petrova P, Tomova R, Ivanov P and Stoyanov S 2021 *Dye. Pigm.* **184**, 108585
[6] Noirbent G and Dumur F 2020 *Europ. Polym. J.* **132** 109702
[7] Grabchev I, Staneva D and Betcheva R 2012 *Cur. Med. Chem.* **19** 4976-4983
[8] GrabchevI, Bosch P, Staneva D 2011 *J. Photochem. Photobiol. A: Chem.* **222** 288-292
[9] Temiz HT, Boyaci IH, Grabchev I and Tamer U 2013 *Spect. Acta Part A: Mol. Biomol. Spectr.*

**116** 339-347
[10] Staneva D, Vasileva-Tonkova E, Bosch P and Grabchev I 2017 *J. Photochem. Photobiol. A: Chem.* **344** 143-148
[11] Li A M and Xu H 2009 *Curr. Med. Chem.* **16** 4797-4813
[12] Ingrassia L, LeFranc F, Kiss R and Mijatovic T 2009 *Curr. Med. Chem.* **16** 1192-1213
[13] Gong H H, Addla D, Lv J S and Zhou C H 2016 *Curr. Top. Med. Chem.* **16** 303-3364
[14] Yu H, Guo Y, Zhu W, Havener K, Zheng X, 2021 *Coord. Chem. Rev.* **444** 2140192
[15] Mlynarczyk DT, Dlugaszewska J, Kaluzna-Mlynarczyk A and Goslinski T 2021 *J. Control Release* **330** 599-617
[16] Sanz del Olmo N, Carloni R, Ortega P, Garcsia-Gallego S and Javier de la Mata F 2020 *Adv. Organomet. Chem.* **74** 1-52
[17] Staneva D and Grabchev I 2021 Chapter 20, Dendrimer as antimicrobial agents, 363-384, Dendrimer-Based Nanotherapeutics, (ed. P. Kesharwani), Elsevier Inc.,
[18] Vasileva-Tonkova E, Grozdanov P, Nikolova I, Staneva D, Bosch P, Medel S and Grabchev I 2018 *Bioint. Res. Appl. Chem.* **8** 3053 - 3059
[19] Vasileva-Tonkova E, Staneva D, Medel S, Bosch P, Grozdanov P, Nikolkova I and Grabchev I 2019 *Fibers Polym* **20** 19-24
[20] Staneva D, Atanasova D, Vasileva-Tonkova E and Grabchev I (in press)
[21] Fischer Mand Georges J 1996 *Chem. Phys. Lett.* **260** 115-118

IOP Conf. Series: Materials Science and Engineering 1188 (2021) 012002 doi:10.1088/1757-899X/1188/1/012002

# Preliminary comparison of contact angle and wetting resistance of water-repellent treated cotton woven fabric

**M Spasova[1], I Rahnev[2]**

[1]Laboratory of Bioactive Polymers, Institute of Polymers, Bulgarian Academy of Sciences, Acad. G. Bonchev St, Bl. 103A, BG-1113 Sofia, Bulgaria

[2] E. Miroglio EAD, BG-8800 Sliven, Bulgaria

mspasova@polymer.bas.bg

**Abstract**. Filtration capability of textile media or the resistance towards the fluid penetration through different fabrics is depending on their surface repellence. The emission of condensed exhaled air on the outer layer of the half facemask from the FFP2 (EN 149) category indicates the termination of its protective properties. The overcoming by the fluid of the textile environment of the mask in its role of an artificial barrier goes through two stages. Initially, the aerosols in the exhaled air condense and meet the repulsion from the inner fabric. Later, due to the hydrodynamic pressure, the condensed droplets pass through the textile medium, overcoming the friction among the fibrous layers. Finally, the droplets irrigate the outer, front layer of the mask and are released into the environment with all the consequences. Closest to this physiological process is the hydrostatic pressure test method. The test of resistance to fluid penetration gives a final assessment of the property of the textile environment without clarifying the elements of the process. Experimental studies have shown similarities in the results of the penetration of the fluid through the textile medium and the initial repulsion of aerosols from the surface of the fibrous layer. The repulsion of water droplets is of predominant importance in relation to the general barrier capacity of the textile media. The main indicator of water repellence is the contact angle of the free drop on the fabric. The subject of this article is the comparison between the contact angle of wetting and the penetration of the fluid by means of standard test methods. The aim of the work is to explain and model the process of fluid penetration through protective masks, as well as to optimize the properties of repulsion and resistance.

**Keywords:** FFP2 masks, COVID-19, fluid filtration, contact angle.

## 1. Introduction

Technological developments in the last 18 months have allowed the identification of two main areas in the concept and production of protective face masks.

During the preliminary achievements in the first direction of the masks, filtering functions were assigned. The technical conditions of these textile garments are given in 149: 2001 + A1: 2009 [1].

At the heart of the design, this type of mask allows airflow without interfering with normal inhalation and exhalation effort. Their functions are aimed at protecting the individual from the environment. These masks as a construction are designed to filter and retain micro particles, including micro-organisms with a size of about 3 microns.

The high requirements for filter masks and the difficult mass production caused the concept of the second direction of the barrier masks. Their technical conditions are given in the specification of the French standardization agency: AFNOR SPEC S76-001 / 2020 [2]. The requirement for these articles is aimed at inhibiting the fluid and associated micro particles on the direction of the directed flow. The purpose of barrier masks is to isolate the individual from the environment and to prevent the spread of infection.

Thus, two directions in the design of the masks were formed: filtration properties by retaining in the volume of the textile medium and repellent properties on the active surface of the medium. From the point of view of the general protection of the masks, their repellent ability has the task of initially and partially slowing down the flow. The study of the repulsion of a fluid with water or oil content presupposes the simultaneous test of air permeability [3], the resistance to wetting [4, 5, 6] and the surface resistance to a water drop. Figure 1 depicts the schematics for the various angles.

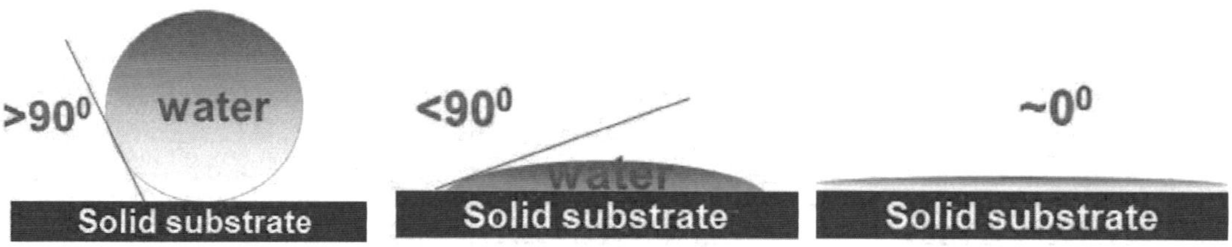


**hydrophobic** **hydrophilic** **superhydrophilic**

**Figure 1**. Visual assessment of a contact angle of fluid on solid substrate.

## 2. Materials and Methods

## 3. Cotton woven fabric

The main subject of the experimental work is a woven fabric with the trade name Filtro. The fibrous composition of the fabric is 100% cotton, the area mass is 157 g/m$^2$. The warp and weft threads are from a single ring yarn: Tt20x1, 100-Cotton. The warp density is 500 threads/dm and the weft density is 245 wefts/dm. The weave is Se3/1Z, and the reverse side with weft effect of the weaving crossings is the active surface of the fabric. In this unequal-faceted fabric structure, the reverse side has a relatively smoother surface than the striated face.

## 4. Fluid repellent treatment

In industrial conditions, the fabric is treated with different chemicals under different technological conditions, such as bath recipe, solution temperature and processing speed. The experimental work follows the algorithm of the factorial experiment. As a result, 8 variants were obtained, which were randomly distributed in the sequence of experiments. The purpose of the treatment is to increase the repellent properties of the fabric surface compared to applied fluids: air, aqueous solutions of mineral salts and oils of natural origin.

After many technological experiments for the water- and oil-repellent properties of the cotton fabric, a recipe with an optimal composition has been determined. The compounds that make up the recipe are based on Dimethyloctadecyl[3-(trimethoxysilyl)propyl]ammonium chloride used as a precursor, PhenOxyEthanol (C6H5OC2H4OH) used as an organic ingredient, as well as the modified silicon

dioxide ingredients. All ingredients of the recipe meet the requirements of OEKOTEX® Standard 100 and REACH, for safe production and application of the processed fabric.
The chemicals are supplied by reputable manufacturers of textile auxiliaries such as Massimo Guarducci Srl, Nano-Care Deutschland AG, Creative Chemical Manufacturer GmbH and Textil Color AG.
The arguments of the factor experiment relate to the temperature of the bath, the immersion time / rate of movement of the fabric and the temperature of drying and polymerization of the thin layer. The temperature of the bath varies from 50°C to 85°C. The speed of the fabric varies from 15 to 19.5 m/min. The drying and polymerization temperatures range from 120 to 160°C. The only peculiarity in the planning of the experiment is the uniformity of the machine mode for the seventh and eighth variants. For the eighth variant, the bath temperature is 65°C, the fabric speed is 18 m/min, the drying temperature is 145°C, but the fabric goes through double immersions in two consecutive bathtubs. This is the reason for getting the best results for the eighth option.

## 5. Characterization

Air permeability was measured using the apparatus DVT HG DLC, DEVOTRANS, Turkey [7], in compliance with the methodological requirements in the standard - Determination of the permeability of fabrics to air, ISO 9237: 1995 [3].

The resistance to water penetration through the fabric was measured with the HIDROTESTER 3H, FX 3000-III, № 851.06.08, TEXTEST AG, Switzerland [8].

Static contact angle measurements of the treated fabrics variants were performed using an Easy Drop DSA20E Krűss GmbH [9] drop shape analysis system (Germany) at 20 ± 0.2 °C. A sessile drop of deionized water with a volume of 10 μL controlled by a computer dosing system was deposited onto the fabric. The contact angles were calculated by computer analysis of the acquired images of the droplet. The data are average from 20 measurements for each sample, (Figure 2).

## 6. Results and discussion

Liquids come in contact with solid surfaces in a variety of applications, including medicine and agriculture. In all kind of applications, knowing the contact angle value gives a strong indication on the properties of the fabricated material. When a drop of water is placed on a solid, it will spread on the surface based on the intermolecular interactions between the solid and the liquid. Water contact angle immediately give an indication of the wettability of the solid surface. If the measured contact angle is above 90 degrees, the solid have poor wetting and is hydrophobic. It is important to determine the water contact angle of the cotton fabric that might be used for the preparation of face masks. The values of the angles of contact for all cotton fabrics were measured with distilled water drop. The representative images of the droplets of the fabric variant 7 were shown in Figure 2. This fabric was almost super hydrophobic with water contact angle of 140.6°±3.4° (Figure 2). The measured values for the other cotton fabrics` variants were hydrophobic with water contact angles ca. 95°, (Table 1).

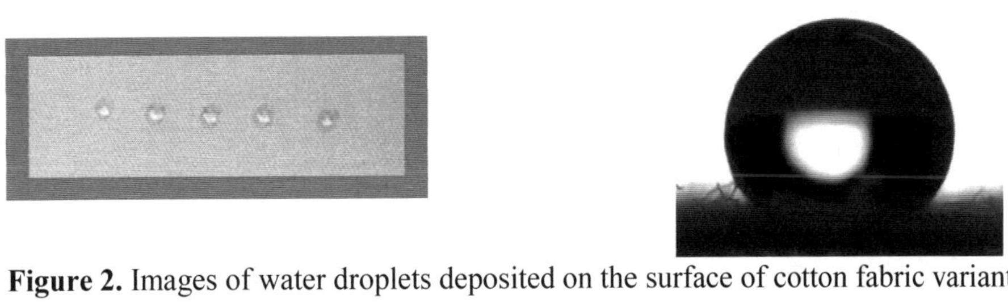

**Figure 2.** Images of water droplets deposited on the surface of cotton fabric variant 7.

Air permeability testing, water penetration tests, spray tests, water repellency and oil repellency test were performed as well. The test results are summarized in Table 1.

**Table 1.** Experimental results on repellent properties of the used cotton fabrics.

| Properties assessment | EN ISO | Dim | Fabric samples variants | | | | | | | |
|---|---|---|---|---|---|---|---|---|---|---|
| | | | 1 | 2 | 3 | 4 | 5 | 6 | 7 | 8 |
| *air permeability* | 9237 | mm/s | 159,7 | 131,9 | 129,1 | 107,5 | 101,7 | 82,8 | 76,7 | 61,1 |
| *Water penetration — Hydrostatic pressure test* | 811 | mmH2O | 108 | 113 | 151 | 190 | 200 | 200 | 215 | 225 |
| *Water repellency* | 14419 | grade | 6 | 5 | 6 | 6 | 5 | 6 | 6 | 5 |
| *Oil repellency* | | grade | 2 | 2 | 5 | 4 | 1 | 4 | 2 | 2 |
| *Static contact angle* | | degree | 84,1 | 94,6 | 95,6 | 96,5 | 99,0 | 102,1 | 105,7 | 140,6 |

The measurement of the surface resistance by the method of the static contact angle gives a relatively more objective assessment than the method of the absorbent drop for a certain period.

In the way of obtaining the contact angle is comparable with the methods for measuring the air permeability or water penetration, due to the use of a measuring device and numerical evaluation of the property.

The convergence of the curves from the measurements of air permeability, water penetration and contact angle is due to the objective approach in the measurement and the detail of the numerical estimates of the indicators.

The significant advantage of the contact angle consists in the possibility of visualization and physical explanation of the acquired water-repellent property.

Figure 3 visualises the graphs of the measured properties for all eight variants.

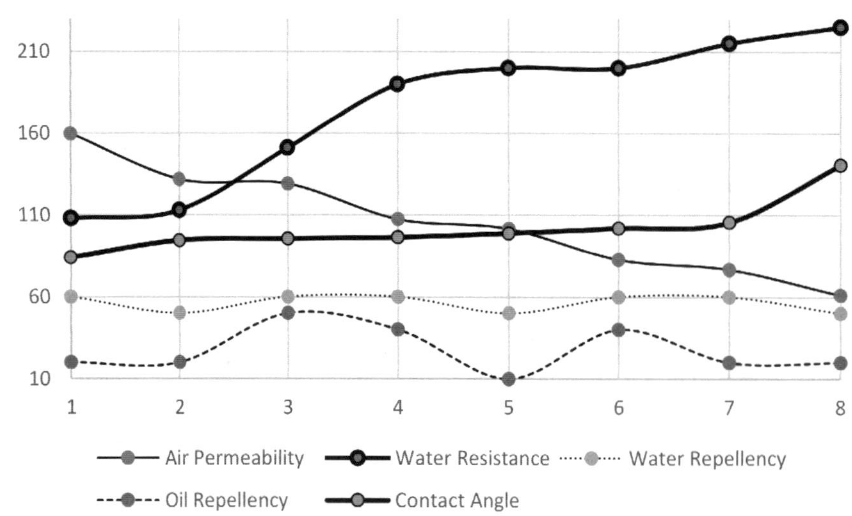


**Figure 3.** Change of the measured indicators by the filtering abilities of the fabric

## 7. Conclusions

Figure 3 clearly illustrates the correctly selected parameters of the technological mode for the presented eight characteristic variants of the factorial experiment. The practical realization of the presented experimental work was preceded by a long period of time in which the trial and error method was accepted as a determining principle. The final version shows the possibility of finding an optimal technological regime in which the necessary water and oil repellent properties of the fabric are achievable.

First of all, the optimization criterion is observed - the balance between air permeability and water penetration resistance. It is seen that the increase of the resistance against the penetration of the water flow is connected with the decrease of the air flow as well. Equilibrium as an optimization criterion consists in maximum filtration / repulsion of the liquefied fluid with minimum resistance to inhalation and exhalation.

Secondly, it was found that the resistance to wetting with a water or oil drop could not be represented by the changes in the 6-point scores shown in the graphs. They have a discrete nature and a very subjective way of measuring. From the graphs shown, no convergence can be established between the variation of water and air resistance, measured with instruments and estimated with direct quantitative indicators, and the corresponding dimension.

In this case, the static contact angle method compensates for the subjective methods and is equivalent in reliability and physical explanation to the hardware measurements.

The method of static contact angle is applicable in the process of technological optimization of fabrics for protective masks. Through the detailed assessment and the possible analysis of the change of the repulsive treatment, the layers in the construction of the mask can be distinguished and the industrial reproducibility can be improved.

**Acknowledgments:** The authors express their gratitude to E. Miroglio EAD for the support and assistance in the implementation of present experimental study.

## References

[1] EN 149:2001+A1, Respiratory protective devices - Filtering half masks to protect against particles – Requirements, testing, marking, https://standards.cen.eu
[2] AFNOR SPEC S76-001/2020, 28.04.2020, Version 1.10, ©AFNOR, www.afnor.org
[3] Textiles — Determination of the permeability of fabrics to air, ISO 9237:1995, https://www.iso.org/standard/16869.html
[4] Textile fabrics - Determination of resistance to surface wetting (spray test), EN ISO 4920 : 2012, www.iso.org
[5] Textiles - Oil repellency - Hydrocarbon resistance test, EN ISO 14419:2010, www.iso.org
[6] Textiles - Determination of resistance to water penetration - Hydrostatic pressure test, EN ISO 811 : 2018, www.iso.org
[7] TOUCH SCREEN AIR PERMEABILITY TESTER DVT HG DLC, https://devotrans.com/en/products/permeability-test-equipment/touch-screen-air-permeability-tester-dvt-hg-dlc.html
[8] FX 3000 HydroTester IV, https://textest.ch/en/portfolio-items/fx-3000-hydrotester-iv/
[9] Y. Gao, S. Jung, L. Pan, Interaction Forces between Water Droplets and Solid Surfaces across Air Films, ACS Omega 2019, 4, 15, 16674–16682.

# Chemical modification and characterization of cotton fabric with 1,8-naphthalimide and its antibacterial activity

**Desislava Staneva[1], Daniela Atanasova[1], Evgenia Vasileva-Tonkova[2], Ivo Grabchev[*3]**

[1]Department of Textile and Leather, University of Chemical Technology and Metallurgy, Sofia, Bulgaria

[2]The Stephan Angeloff Institute of Microbiology, Bulgarian Academy of Sciences, Sofia, Bulgaria

[3]Faculty of Medicine, Sofia University, "St. Kliment Ohridski", Sofia, Bulgaria

i.grabchev@chem.uni-sofia.bg

**Abstract**. 1,8 naphthalimide (NI) containing pyridinium nuclei attached by a methylene bridge to the imide nitrogen atom or at the C-4 atom has been obtained by a single-step synthesis. The main photophysical characteristics of NI in organic solvents and water have been studied. A cotton fabric has been modified with chloroacetyl chloride. The possibility of chemical binding of NI to the modified cotton fabric, by quaternization of the pyridinium nuclei and by physical deposition and retention of NI to the surface of the cotton fabric was investigated. The antimicrobial test showed good activity of NI against Gram-positive and Gram-negative bacteria and a fungal strain of the genus Candida, which is preserved after its deposition on cotton fabric. It has been shown that covalent bonding contributed for a better activity compared to the physical deposition and retention of NI on the surface of the cotton fabric.

## 1. Introduction

In recent years, there has been an increased resistance of microorganisms to antibiotics used in clinical practice [1,2]. Despite the wide range of modern antibiotics and new approaches to combating this resistance, recurrent infections remain a problem for human health and quality of life. This requires searching for and finding effective compounds exhibiting antibacterial activity.

1,8-naphthalimide are a special class of heterocyclic fluorophores that emit blue or yellow-green fluorescence. They are used as dyes for polymeric materials and textiles [3], in the design of sensor systems for the detection of metal ions and protons [4]. Also some 1,8-naphthalimides derivatives of are of the wide biomedical interest due to their high antibacterial, antifungal, anti-inflammatory, antiviral, or antitumor activity [5-7]. The color and fluorescence characteristics of the 1,8-naphthalimides depend on the polarity of the chromophore system. In the presence of electron donor substituents, such as amino derivatives, their color is yellow, which is further enhanced by the emitted yellow-green fluorescence. This color is preserved even after their deposition on various textile or polymeric materials which makes them suitable for dyeing in order to give them an intense color [3].

It is of great interest to combine the properties of some 1,8-naphthalimides, which have shown good microbiological activity with their ability to dye textile materials, thus handing them antibacterial or antifungal activity [8-10].

The aim of this study was to synthesize a 1,8-naphthalimide with pyridinium units able to form chemical bonds with modified cotton fabric with chloroacetyl chloride. The antimicrobial activity against model Gram-positive and Gram-negative bacteria and yeasts in solution and after deposition on a cotton fabric has also been investigated.

## 2. Experimental part

### *2.1. Materials and methods*

Thermo Spectronic Unicam UV 500 spectrophotometer and "Cary Eclipse" spectrofluorometer have been used for photopysycal measurements. The spectra were recorded using a 1 cm path length synthetic quartz glass cell at concentration $10^{-5}$ mol $L^{-1}$. The fluorescence quantum yield ($\Phi_F$) has been calculated on the basis of the absorption and fluorescence spectra by equation 1.

$$\Phi_F = \Phi_{st} \frac{S_u}{S_{st}} \frac{A_{st}}{A_u} \frac{n_{Du}^2}{n_{Dst}^2} \quad (1)$$

where the $\Phi_F$ is the emission quantum yield of the sample; $\Phi_{st}$ is the quantum yield of the standard Rhodamine ($\Phi_{ref}$= 0.94 [11]); $A_{st}$ and $A_u$ are the absorbance of the standard and sample at the excited wavelength, respectively; while $S_{st}$ and $S_u$ are the integrated emission band areas of the standard and sample, at the excited wavelength respectively, and $n_{st}$ and $n_u$ is the solvent refractive index of the standard and sample; subscripts *u* and *s* refer to the unknown and standard, respectively. All organic solvents are of spectroscopy grade: dioxane; tetrahydrofurane (THF); chloroform; acetonitrile; N,N-dimethylformamide (DMF); dimethylsulfoxide (DMSO); ethanol (EtOH); methanol (MeOH); water. 1H NMR (600.13 MHz) and 13C (150.92 MHz) spectra were acquired on an AVANCE AV600 II+ NMR spectrometer. Fourier-transform spectrometer (IRAffinity-1 Shimadzu, Kyoto, Japan) at a 2 $cm^{-1}$ resolution. The color coordinates (L*a*b*, XYZ and xy) of dyed cotton fabrics have been determined by using Datacolor Spectraflash SF300 spectrophotometer (Datacolor, NJ, USA). The bifunctional chloroacetyl chloride (CAC) has been used without further purification as obtained from Sigma-Aldrich.

### *2.2. Synthesis of 2-(pyridin-4-ylmethyl)-6-((pyridin-4-ylmethyl)amino)-1H-benzo[de] isoquinoline-1,3(2H)-dione (NI)*

4-nitro-1,8-naphthalc anhydride (2.43 g, 0.01mol) was dissolved in 50 ml DMF and 4-(aminomethyl)pyridine (2,5 ml, 0.025 mol) was added. The mixture was stirred 6 hours, at 80°C then poured into 200 ml of water. The precipitate was filtered, washed with water and dried under vacuum. Yield: 3,51g, 89 %;

FT-IR (KBr) $cm^{-1}$: 3274, 3064, 2967, 2837, 2790, 1674, 1636, 1581, 1551, 1386, 1364, 1340, 1244, 1123, 1018, 775, 751; $^1$H NMR ($CDCl_3$, δ, ppm): 8.55 (d; J=7.36 Hz; 4H); 8.48 (d; J=8.22 Hz; 1H); 8.26 (d; J=8.35 Hz, 1H); 8.04. (t, J=6.6 Hz; 1H); 7.40 (d, J=8.09 Hz; 4H); 7.26 (d, J=8.11 Hz; 1H); 4.44 (s; 1H); 4.30 (s; 1H); $^{13}$C NMR ($CDCl_3$, δ, ppm): 163.7, 163.1, 150.3, 145.3, 144.1, 136.3, 131,2, 130.3, 128.4, 127.4, 125.9, 113.3, 110.0, 45.1,

Elemental analysis: $C_{24}H_{18}N_4O_2$ (394.21 g $mol^{-1}$): Calc. (%): C 73.06, H 6.20, N 13.03; Found (%): C 70.43, H 6.13, N 13.09.

### *2.3. Modification of cotton fabric with chloroacetyl chloride [10]*

1.0 g of cotton fabric (140 g / $m^2$) is placed in a beaker and soaked in DMF. At room temperature for 30 minutes, chloroacetyl chloride 10% (v/v) dissolved in DMF was added and stirred for 90 minutes, after which the cotton fabric has been removed, washed thoroughly with water and dried. The solution was of a liquor to goods ratio of 20:1.

*2.4. Dyeing of cotton fabric with NI (Cotton1-NI)*
1.0 g of cotton fabric (140 g / $m^2$) was immersed into the NI (5 mg dissolved in 10 ml DMF) at 25°C. After dyeing for 30 min, the sample was washed with water and dried at room temperature

*2.5. Dyeing of modified cotton fabric with NI (Cotton2-NI)*
1.0 g of the modified cotton fabric has been placed in a flask and 20 ml of DMF was added. Heat to 80°C and over 30 minutes and a solution of 5 mg of NI dissolved in 5 ml of DMF was added. Under these conditions for 6 hours has been stirred, then the colored cotton fabric has been washed thoroughly with water. The unreacted dye is removed by the boiling of cotton fabric in ethanol. All future measurements and tests were performed on this sample.

*2.6. Antimicrobial activity test*

*2.6.1. Minimum Inhibitory Concentration (MIC)*
Broth dilution test was used for determination of the MICs of NI against Gram (+) *B. cereus* and Gram (-) *P. aeruginosa* as test cultures. The compound was serially diluted in tubes with meat-peptone broth (MPB) in concentrations in the range 20 - 1000 μg/ml. Control tubes without added compounds were also prepared for each microbial culture. The tubes were inoculated with respective microbial suspension and incubated at appropriate temperature for 24 h. Microbial growth was assessed by the turbidity at 600 nm ($OD_{600}$). The growth control, sterility control and sample control were used. The lowest concentration of the samples that inhibited the visible growth of the strains was referred as MIC. Three independent experiments were carried out and averages are given.

*2.6.2. Antibacterial activity of modified cotton fabrics*
Modified with NI cotton fabrics was tested for antibacterial activity against *B. cereus* and *P. aeruginosa* as model strains. Test tubes containing MPB and square cotton speciments (1 cm x 1 cm, 25 mg) were inoculated with cell suspension of each bacterial culture. Tubes with untreated cotton and without speciments were also prepared as controls. After incubation for 24h at appropriate temperature, the specimens were removed and $OD_{600}$ was determined. The antimicrobial activity of the treated cotton samples was evaluated by the reduction of $OD_{600}$ after incubation compared to the control sample. All antimicrobial tests were done in triplicate and the average was taken; standard deviations were less than 5%.

## 3. Results and discussion

1,8-naphthalimide (NI) has been obtained in a single-step synthetic route as shown in Scheme 1. 4-Nitronaphthalic anhydride was reacted with 4-(aminomethyl)pyridine in a DMF medium at 80°C to simultaneously nucleophilically replacement the nitro group at C -4 with 4-(aminomethyl)pyridine and condensation of the primary amino group in the anhydride ring. Thus, 1,8-naphthalimide containing two pyridinium nuclei has been synthesised. The choice of these substituents has been dictated by the fact that the pyridinium core can be quaternized with the residues of the acylated cotton fabric, and thus to form a covalent bond of NI with the cotton matrix.

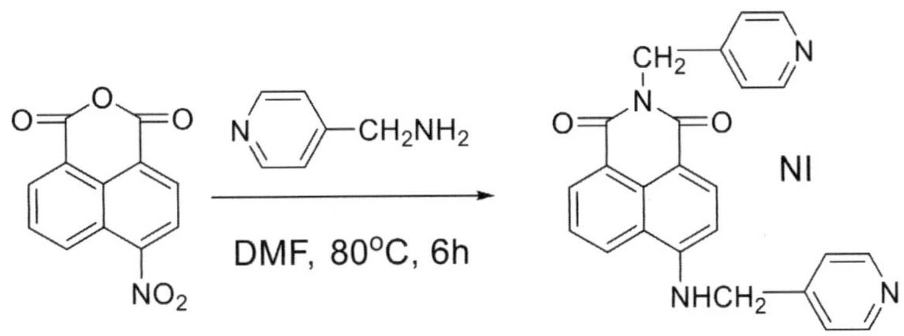


Scheme 1. Synthesis of 1,8-naphthalimide NI

In order to obtain a yellow colored cotton fabric with microbiological activity, 1,8-naphthalimide, has been used as dye using two approaches. In the first method, unmodified cotton fabric was used, and NI was physically deposited by irrigation at room temperature for 30 minutes, then washed with water and dried. In this way, NI is deposited on the cotton fabric, and the way it is retained on the surface is mainly through hydrogen bonds and van der Waals forces of attraction. Poor solubility in the water further aids its retention (Cotton1-NI).

In the second case, a pre-modified with chloroacetyl chloride cotton fabric has used [10]. It has been used due to the different degree of activity of the two chlorine atoms. The much more reactive acyl chlorine atom reacts with the hydroxyl groups of the cotton, leaving a second reactive chlorine atom. 1,8-naphthalimide under study has two pyridinium nuclei in its structure that can participate in a quaternization reaction with alkyl chlorine substituents (Scheme 2). The quaternization reaction has performed under DMF at 80 °C for 6 hours. After boiled in ethanol for 30 minutes, the color of the fabric remains yellow, which means that NI has covalently bound to the cotton fabric (Cotton2–NI).

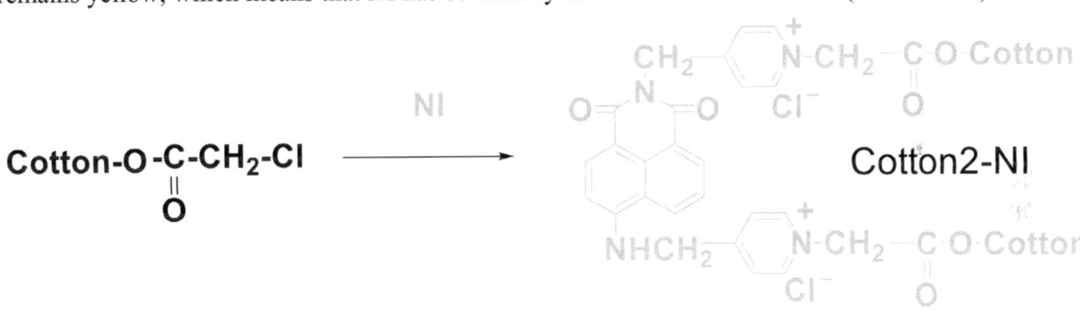


Scheme 2. Chemical modification of cotton fabric with NI

The basic photophysical characteristics of NI have been studied in organic solvents with different polarities and water (Table 1). NI is yellow in the solvents tested. Its absorption maxima are in the spectral range $\lambda = 412\text{-}436$ nm, and the emitted fluorescence has maxima at $\lambda = 495\text{-}536$ nm. Figure 1 shows the dependence of the position of these maxima on the polarity of the medium (ET30). The results show that as the polarity of the solvents increases, the absorption and fluorescence maxima are shifted batochromically, and NI exhibits positive solvatochromism. The Stokes shift ($\nu_A$-$\nu_F$) values are in the range 3730-4170 $cm^{-1}$, which are typical of 1,8-naphthalimide derivatives with alkylamino substituents at C-4 [12,13].

Table 1. Photophysical characteristics of NI in organic solvents and water

| Solvents | $\lambda_A$ nm | $\lambda_F$ nm | $\nu_A$-$\nu_F$ $cm^{-1}$ | $\varepsilon$ mol l $cm^{-1}$ | $\Phi_F$ |
|---|---|---|---|---|---|
| DMF | 432 | 515 | 3730 | 8400 | 0.26 |
| Water | 438 | 536 | 4170 | 8160 | 0.12 |

| | | | | | |
|---|---|---|---|---|---|
| Etanol | 437 | 519 | 3620 | 9790 | 0.32 |
| Dioxane | 412 | 495 | 4070 | 8460 | 0.32 |
| Methanol | 437 | 523 | 3760 | 8550 | 0.25 |
| Chloroform | 415 | 501 | 4140 | 8450 | 0.30 |
| DMSO | 435 | 520 | 3760 | 7440 | 0.24 |
| THF | 421 | 498 | 3670 | 8210 | 0.27 |
| Acetonitrile | 422 | 509 | 4050 | 8610 | 0.28 |

The quantum efficiency of NI has been determined by calculating the quantum yield of fluorescence. The data in Table 1 show that for organic solvents the PF is in the range $\Phi_F = 0.24$-$0.32$, which means that the polarity of the medium does not significantly affect this indicator. A significantly lower value was obtained in aqueous solution ($\Phi_F = 0.12$). In this case, in addition to the polarity of the medium, the dipole-dipole interaction between water molecules and NI also has an effect, which leads to a decrease in the polarization of the chromophore structure.

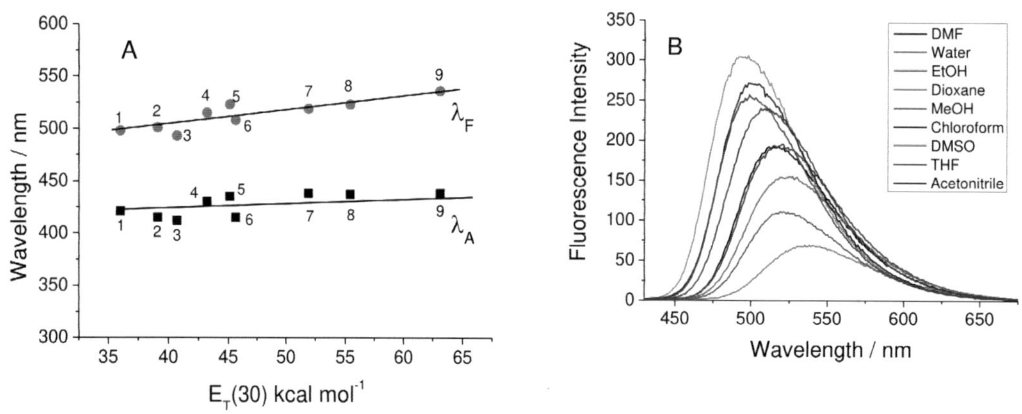


Fig. 1. Dependence of absorption and fluorescence maxima of NI2 on the empirical parameter of solvent polarity $E_T(30)$: 1-dioxane; 2-THF; 3-chloroform; 4-acetonitrile; 5-DMF; 6-DMSO; 7-EtOH; 8-MeOH; 9-water (A) and fluorescence spectra of NI at the same solvents (B).

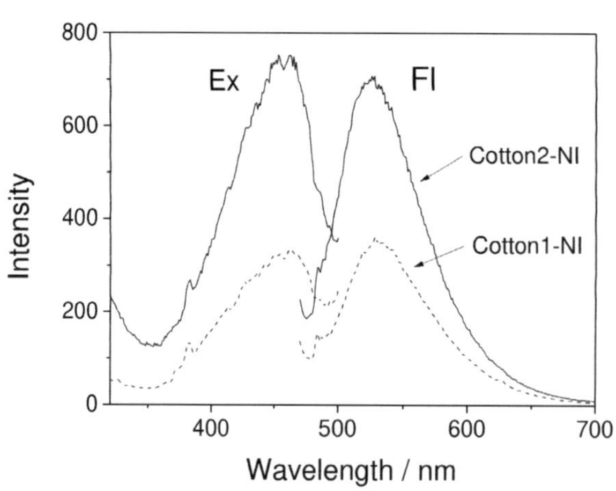


Figure 2. Excitation (Ex) and fluorescence (Fl) spectra of cotton fabrics

The excitation and fluorescence spectra of the colored cotton fabrics obtained by both methods were studied (Figure 3) and the corresponding maxima were determined. It was found that at Cotton1–NI the maxima are at 462 nm and 527 nm, while in Cotton2–NI they are at 456 nm and 527 nm, respectively, but the fluorescence intensity is almost twice high. This difference can be explained by the quaternization of the nitrogen atoms of the pyridinium structure in the formation of the chemical bond with the cotton fabric.

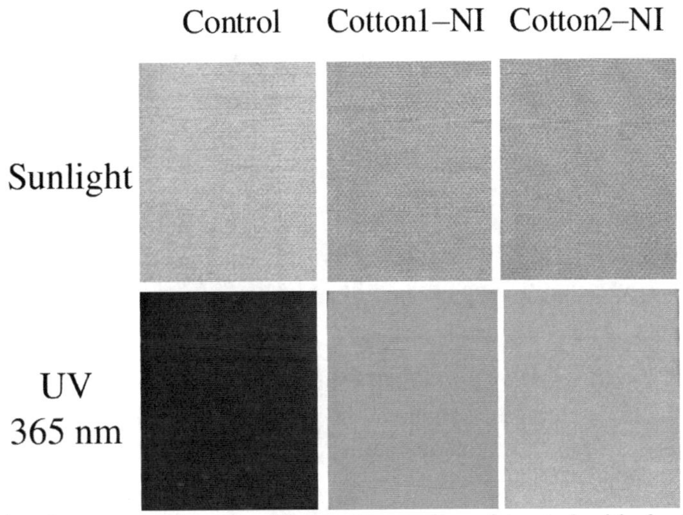


Figure 3. Micrograph of non-treated cotton fabrics (Control) and treated with dendrimer Cotton1–NI and Cotton2–NI upon their irradiation with Sunlight and UV light

Figure 3 shows the change in the color of Cotton1–NI and Cotton2–NI cotton fabrics compared to that of the untreated cotton fabric during the irradiation with UV light (λmax = 365 nm) and daylight. The untreated cotton fabric has a white color in daylight. Also It is seen that, when irradiated with daylight the treated cotton fabrics have a saturated yellow color, while when irradiated with UV light the color changes to yellow-green.

The chromaticity coordinates (x and y) and CIEab coordinates (L*, a* and b*) of both cotton fabric has been determined by Datacolor systems using equations 2-4 [14] :

$$L^* = 116\,(Y/Y_0)^{1/3} - 16 \quad (2)$$
$$a^* = 500\,[(X/X_0)^{1/3} - (Y/Y_0)^{1/3}] \quad (3)$$
$$b^* = 200\,[(Y/Y_0)^{1/3} - (Z/Z_0)^{1/3}] \quad (4)$$

were the tristimulus values of specified achromatic light used in illumination are $X_0$, $Y_0$, $Z_0$, The values of $Y_0$ is normalized in such a way that $Y_0$ = 100 and the respective X, Y, Z coordinates has been determinate. As a standard has been used a non-treated cotton fabric. The date obtained are summarised in Table 2. The results show that the dyed cotton fabrics have a yellow color.

Table 2. Color characteristics of non-treated cotton fabric and treated with NI

| | L* | a* | b* | X | Y | Z | x | y |
|---|---|---|---|---|---|---|---|---|
| Non treated cotton | 93.77 | -0.26 | 3.75 | 80.22 | 84.75 | 85.65 | 0.3201 | 0.3382 |
| Cotton1–NI | 86.76 | 0.16 | 24.09 | 65.99 | 69.52 | 48.13 | 0.3593 | 0.3786 |
| Cotton2–NI | 88.07 | -3.11 | 46.06 | 67.06 | 72.22 | 31.83 | 0.3919 | 0.4221 |

The results show that the way of connecting the NI to the cotton matrix affects the color characteristics of the fabric. In both cases the color of the cotton fabric is yellow but with a different shade. In covalent bonding, it has a deeper color but is more brilliant due to the emission of the more intense fluorescence. This difference can be explained by the quaternization of the pyridinium nuclei, as that is in the C-4 position will have a greater influence.

The antimicrobial activity of NI has been investigated against Gram-positive bacteria *B. cereus*, Gram-negative bacteria *P. aeruginosa*, and the fungus *C. lipolytica*. The MICs of the compounds against the model strains have been determined in MPB. The inhibition of the growth of Gram-positive *B. cereus* was higher (the lowest MIC, 250 mg /ml; 0.63 mM) while in *C. lipolytica* the activity is approximately two-fold lower (MIC - 1,40 mM). Significantly lower activity has been determined against Gram-negative *P. aeruginosa* ($MIC_{50}$ at 1.01 mM). Based on classification of Holetz et. al. antimicrobial activity of the investigated compounds can be considered moderate [15].

Various methods have been proposed for the antimicrobial modification of material surfaces as an alternative way of preventing the formation of highly resistant biofilms [16]. The antimicrobial effect of the treated with NI cotton fabrics has been evaluated by the reduction of the growth of the model bacteria Gram-positive *B. cereus* and Gram-negative *P. aeruginosa* bacteria. Against *P. aeruginosa,* both cotton samples showed similar reduction of the growth (5-10%) (Figure 4). The low activity in this case is due to the fact that NI is firmly attached to the cotton surface, which does not allow to diffuse from the fabric into the solution. This significantly reduces the possibility of direct contact of bacteria with NI. Against *B. cereus*, the activity was better expressed in Cotton2-NI (30%) and 10% in Cotton1-NI. Probably in this case the difference is due to the presence of quaternary groups, which more easily damage the membrane of Gram-positive bacterial cells.

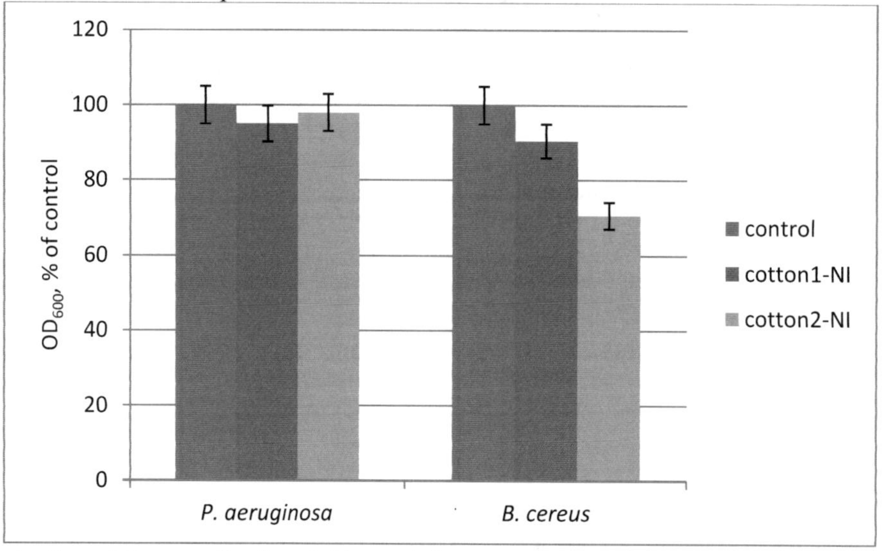


Figure 4. Reduction of the growth of model bacteria by treated cotton fabrics

## 4. Conclusion

By one-step synthesis, 1,8-naphthalimide has been obtained, containing pyridinium nucleattached to the imide nitrogen atom and to C-4 via methylene (-CH2-) spacer. A cotton fabric was dyed by two methods, in which NI is retained on the surface of the fabric by physical forces of attraction or by forming a strong covalent bond. In the second case, the cotton fabric is modified with chloroacetyl chloride and then it interacts with NI. The color parameters of both cotton fabrics were determined and it was shown that a brighter yellow color is obtained with covalently bonded NI. The photophysical characteristics of NI in various organic solvents have been studied and it has been shown that in solution it has an intense yellow color and fluorescence, which are preserved even after its application on a cotton fabric. The

IOP Conf. Series: Materials Science and Engineering 1188 (2021) 012003 doi:10.1088/1757-899X/1188/1/012003

fluorescence intensity is almost twice as high in the case where the NI is covalently bonded to the cotton fabric. NI showed moderate activity against the tested cultures, better expressed against Gram-positive bacteria. The antibacterial activity of the stained fabrics was studied and found to be more pronounced in covalent binding of NI.

**Acknowledgments**
This work was supported by Grants КП-06-КОСТ/19, Fund Scientific Research, Ministry of Education and Science of Bulgaria.

**References**

[1] Carl N and Otto C *N Engl. J. Med.* 2014, **371**, 1761-1763
[2] Yagui M 2018 *Rev. Peru Med. Exp. Salud Publica* **35** 7-8
[3] Dodangeh M, Grabchev I, Staneva D and Gharanjig K 2021 *Fibers. Polym.*
[4] Panchenko P A, Fedorova O A and Fedorov Yu V 2014 *Russ. Chem. Rev.* **83** 155–182
[5] Li A M and Xu H 2009 *Curr. Med. Chem.* **16** 4797–4813
[6] Ingrassia L, LeFranc F, Kiss R and Mijatovic T 2009 *Curr. Med. Chem.* **16** 1192–1213
[7] Gong H H, Addla D, Lv J S and Zhou C H 2016 *Curr. Top. Med. Chem.* **16** 303–3364
[8] Vasileva-Tonkova E, Grozdanov P, Nikolova I, Staneva D, Bosch P, Medel S and Grabchev I, 2018 *Biointerface Research in Applied Chemistry* **8 (1)** 3053 – 3059
[9] Cangiotti M, Staneva D, Ottaviani M F, Vasileva-Tonkova E and Grabchev I 2021 *J. Photochem Photobiol, A: Chemistry.* **415** 113312
[10] Staneva D, Vasileva-Tonkova E and Grabchev I 2019 *J. Photochem. Photobiol. A: Chemistry* **382** 111924
[11] Fischer M and Georges J 1996 *Chem. Phys. Lett.* **260** 115-118
[12] Yordanova-Tomova S, Cheshmedzhieva D, Stoyanov S, Dudev T, and Grabchev I 2020 *Sensors (Switzerland)* **20 (14)**, art. no. 3892, 1-14
[13] Grabchev I 1998 *Dyes Pigm*, **38**, 219-226
[14] B. Becerir, 2003 *Colourage* **50**, 39–48
[15] Adlhart C, Verran J, Azevedo N.F., Olmez H., Keinänen-Toivola M.M., Gouveia I., Melo L.F., Crijns F. 2018 *J. Hosp. Infect.* **99(3)**, 239-249
[16] Holetz F B, Pessini G L, Sanches N R, Cortez D A G, Nakamura C V and Filho B P D, 2002 *Mem. Inst. Oswaldo Cruz.* **97** 1027-1031

IOP Conf. Series: Materials Science and Engineering 1188 (2021) 012004 doi:10.1088/1757-899X/1188/1/012004

# Modified with chitosan cotton fabric for control release of indomethacin

**Daniela Atanasova[1], Desislava Staneva[1], Ivo Grabchev[2]**

[1]Department of Textile and Leather, University of Chemical Technology and Metallurgy, Sofia, Bulgaria

[2]Faculty of Medicine, Sofia University, "St. Kliment Ohridski", Sofia, Bulgaria

grabcheva@mail.bg

**Abstract**. The study aims to obtain and characterize new composite materials with potential application as antimicrobial wound dressings releasing an anti-inflammatory biologically active substance indomethacin. The preparation of the material involves an impregnation of the cotton fabric with chitosan crosslinked with citric acid. Indomethacin interacts with the chitosan by ionic bonds at the appropriate pH. The obtained materials were characterized by infrared spectroscopy and optical microscopy. The mechanism of the indomethacin release from composite materials in phosphate buffer pH 7.4 at 37 °C was examined using different kinetics models. The concentration of citric acid affects the release of the drug, and the material with the highest concentration of citric acid has more antimicrobial activity than the other samples. The composites showed a slightly higher antimicrobial effect against Gram-negative *Pseudomonas aeruginosa* than against Gram-positive *Bacillus cereus*.

## 1. Introduction

Textile materials modified with high molecular weight biologically active substances find application in medical practice in the local treatment of the skin and transdermal therapy [1]. One of the advantages of these materials is their active interaction with the environment [2]. They can react to changes in external conditions with controlled biomedical action (release of biologically active substances, absorption of substances released during sweating, sensor properties, antimicrobial activity, etc.) [3].

In recent years, polysaccharides are a subject of intensive study due to their numerous advantages: biocompatibility, biodegradability, good permeability, wide distribution in nature, low cost and others [4,5]. Starch, carboxymethylcellulose, alginate, carrageenan, and chitosan are polysaccharides utilize in textile processing. They are also used for hydrogel preparation with a biomedical application [6].

Chitosan has many applications in the textile industry, such as wastewater treatment, odour removal, antistatic, and crease-resistant finishing. It also enhances textile dyeability, printability, and increase the obtained colour strength. In the medical textile, chitosan finds application as wound dressing and sutures as it accelerates wound healing, blood coagulation and has antimicrobial properties. [7,8].

Textile materials modified with chitosan hydrogel are suitable composites for the gradual release of a biologically active substance. Chitosan is a natural stimuli-responsive polymer that is capable of showing both thermo-responsive and pH-responsive properties. The presence of free amino and hydroxyl groups in its structure allows it to be modified and interact with various inorganic and organic compounds [9].

Usually, the chitosan dissolved in a dilute solution of acetic acid has applied to the treatment of cotton fabric. However, this method does not create any strong chemical bonds between chitosan and cotton fabrics. As the chitosan is water-soluble, the obtained modification of fabric is not durable in repeatable washing. Therefore, it is necessary to use a crosslinking agent to link the reactive groups of chitosan and cellulose to improve the durability and efficiency of the resulting material [10]. Citric acid can interact with the hydroxyl functional groups of chitosan and cellulose and form covalent bonds with heat treatment and in the presence of catalyst [11]. Another possible interaction is through ionic bonds between protonated amine groups and deprotonated carboxylate ions [12].

The chitosan-treated cotton fabrics have antimicrobial properties when the amino groups of chitosan are ammonium salts in dilute acid solution. The salt then binds to the negatively charged surface of the microorganism. Protonated amino groups of chitosan can also form ionic bonds with the groups of biologically active substances. This pH-dependent interaction and subsequent separation is a way to obtain controlled drug release.

Indomethacin is a non-steroidal anti-inflammatory drug. It acts by blocking in the body an enzyme called cyclooxygenase that is involved in the production of irritating chemicals in response to surgery or illness [13]. Its controlled release has shown the potential to accelerate wound healing and the inflammatory processes [14,15]. However, its disadvantage is its low solubility in water. Therefore, it is not enough for the hydrogel alone to swell to release the drug, but the process of its release can be aid by hydrogel erosion. Indomethacin is acidic due to the ionization of the carboxylic acid group and has a pKa value of 4.5. The molecules of indomethacin and chitosan can interact with electrostatic bonds, which association and dissociation are pH-dependent.

The study aims to obtain new composite materials with potential application as antimicrobial dressings for wounds that released an anti-inflammatory biologically active substance (indomethacin). The cotton fabric was modified with chitosan, indomethacin and citric acid bonded via electrostatic bonds. The influence of citric acid on the indomethacin release and antimicrobial activity of the composite materials were investigated.

## 2. Materials and methods

### 2.1. *Materials*

Chitosan with molecular weight ranging from 600 000 to 800 000 was purchased from Acros organics. Citric acid, indomethacin and glacial acetic acid were used as obtained from Sigma Aldrich. A cotton fabric with a surface weight of 135±5 g/m$^2$ was used throughout the work. Sodium dihydrogen phosphate dehydrate and di-sodium hydrogen phosphate dodecahydrate were obtained from Merck. All solutions were made with distilled water.

### 2.2. *Preparation of composite materials*

Chitosan (2.7% w/v) was dissolved in water with stirring, gradually adding glacial acetic acid (1% v/v) till obtaining a clear, viscous solution. The solution was then left at room temperature for 24 hours to remove air bubbles formed during stirring. The indomethacin was dissolved in ethanol and the corresponding quantity (26 mg/g chitosan) was added to the chitosan solution. The pH of the solution was measured as pH 5.5. The cotton fabric was impregnated with the solution of the citric acid in different concentrations (1%, 5%, and 10% w/w chitosan) and dried at room temperature. The obtained samples were named m1, m2, and m3, respectively. Next, the samples were impregnated with the solution of chitosan and indomethacin. The volume of solutions was of liquor to goods ratio 2:1. The samples were dried at 50 °C for 40 minutes.

### *2.3. Analysis*

The cotton fabric and the composite material m3 were analyzed by Optical microscope Levenhuk Rainbow D50L PLUS 2M Digital Microscope, Moonstone with 2 Mpx Digital Camera. The magnification of the pictures was x40. IR analysis was carried out on an Infrared Fourier Transform spectrometer (IRAffinity-1, Shimadzu) equipped with a diffuse-reflectance attachment (MIRacle Attenuated Total Reflectance Attachment). Measurements were done using spectral range 600–4000 $cm^{-1}$. The pH of the prepared solution was measured with Hanna instruments, Microprocessor pH Meter. The release of the indomethacin from composite materials was followed with ONDA Spectrophotometer UV-31 SCAN, 190-1100 nm.

### *2.4. In Vitro Drug Release Studies*

The indomethacin release has been following in pH 7.4, 0.01 M phosphate buffer at 37± 0.5 °C by measuring the absorbance at λ=320 nm. Each sample was irrigated at 20 ml buffer. 5 mL of solution was withdrawn from the vessel and replaced with 5 mL new amount of phosphate buffer at 37 °C. The concentration of withdrawn solution was determined by UV−vis spectrometry at 320 nm against a standard curve obtained by measuring indomethacin solutions of known concentrations. Each determination was carried out in triplicate.

To identify the mechanism for the release of indomethacin from crosslinked chitosan, the suitability of four kinetics models were evaluated according to the equations presented in Table 1. $Q_t$ is the cumulative amount of drug release at time t. $Q_o$ is the initial amount of drug in the solution. K is the release constant of the relevant model. $M_t/M_\infty$ is a fraction of drug release at time t. n is an exponent. The value of n characterized the release mechanism of the drug: $0.45 \leq n$ corresponds to a Fickian diffusion mechanism; $0.45 < n < 0.89$ to non-Fickian transport, n=0.89 to Case II (relaxational) transport; n>0.89 to super Case II transport.

Table 1. Kinetic models for drug release

| Kinetic model | Equation |
|---|---|
| Zero order | $Q_t = Q_o - kt$ |
| First order | $LogQ_t = LogQ_o - kt/2.303$ |
| Higuchi | $Q = kt^{1/2}$ |
| Korsmeyer-Peppas | $M_t/M_\infty = kt^n$ |

### *2.5. Antimicrobial assay of the treated cotton fabrics*

Agar diffusion assay was firstly used for testing the antimicrobial activity of the treated cotton fabrics against Gram-positive *B. cereus* and Gram-negative *P. aeruginosa* as model bacterial strains. Petri plates with Mueller-Hinton agar were seeded with aliquots of cell suspensions of the test cultures. Cotton speciments (10 mm x 10 mm) were placed onto the agar surface and the plates were incubated at appropriate temperature and microbial growth was monitored for 24 h. The antimicrobial activity was indicated by the presence of clear zones around the speciments.

The antimicrobial effect of the treated cotton fabrics has also been tested in meat-peptone broth (MPB) by the reduction of the growth of the model strains. Tubes containing sterile MPB and square cotton speciments (10 mm x 10 mm) were inoculated with each bacterial suspension. Tubes with untreated cotton samples and without speciments were also prepared as controls. After incubation for 24 h at appropriate temperature, microbial growth was assessed by measuring the optical density at 600 nm ($OD_{600}$). The antimicrobial activity of the treated cotton samples was evaluated by the reduction in cell density after incubation. All antimicrobial tests were done in triplicate.

IOP Conf. Series: Materials Science and Engineering 1188 (2021) 012004 doi:10.1088/1757-899X/1188/1/012004

## 3. Results and discussion

Chitosan is a polyelectrolyte with cationic charged amino groups in a water solution under pH 6.5. The ability of chitosan-based drug carriers to sorb and release active substances is influenced by their preparation conditions involved in the crosslinking process, such as the ratio of chitosan to crosslinker, pH, and temperature. Citric acid is a weak polycarboxylic acid. It is suitable as a crosslinking agent for polysaccharides as chitosan. The deprotonated carboxylic groups of citric acid can form ionic bonds with the ammonium groups of chitosan, as has shown in Figure 1. Indomethacin has a pKa of 4.5, and at pH above the pKa, its molecules dissociate into ionized form. In the pH range 4.5 - 6.5, the electrostatic interaction is possible between negatively charged indomethacin and positively charged chitosan [16]. The indomethacin and chitosan interaction, as well the chitosan and citric acid crosslinking, depend on pH.

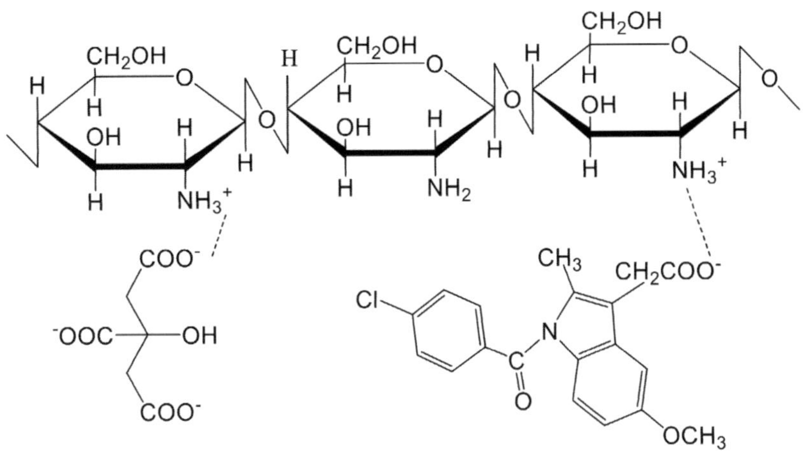

Fig. 1. Interaction of chitosan with citric acid and with indomethacin

Figure 2 presents the photographs taken with a digital camera under optical microscope observation of the cotton fabric and composite material m3, obtained with the highest quantity of citric acid. The chitosan layer covers the fibre surface and glues some of them together.

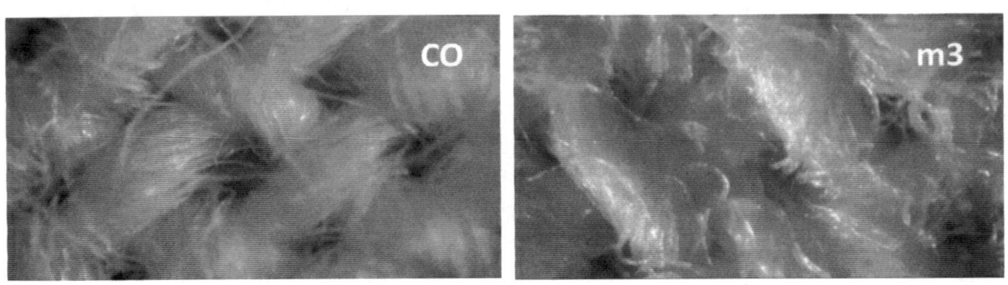

Fig. 2. Optical microscope photographs of cotton fabric and composite materials m3 (magnification of the pictures is x 40)

FTIR experiments were performed to study the interaction between chitosan and citric acid. The FTIR spectra of the cotton fabric and samples m1, m2 and m3 are shown in Figure 3. The main component, cotton, has approximately the same functional groups as the chitosan. The quantity of the citric acid and the indomethacin is small, and this complicated the FTIR analysis. The tool for manage this problem is spectral subtraction of the main component, cotton fabric, from the others. It makes it easy to see and discuss peaks of citric acid and indomethacin.

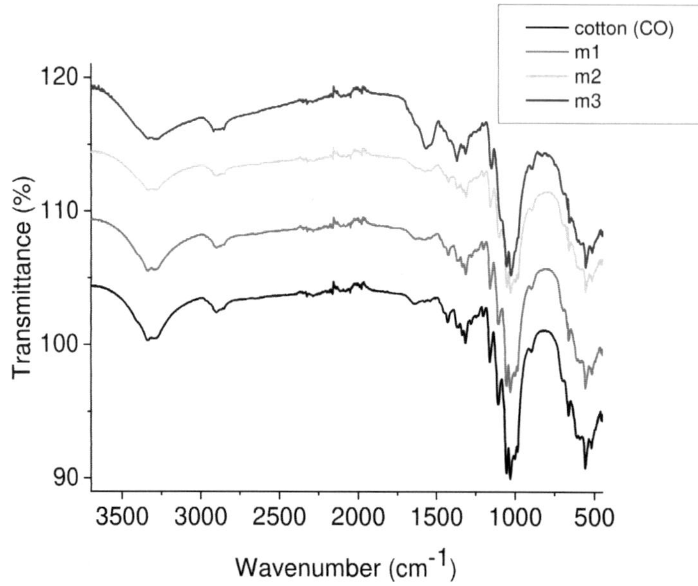


Fig. 3. FTIR spectra of cotton fabric (co) and composite materials (m1, m2, and m3)

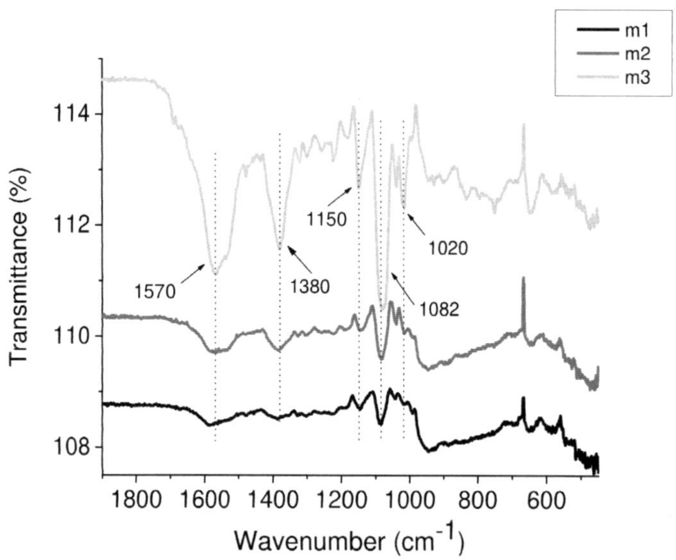


Fig. 4. Substracting FTIR spectrum of cotton fabric from the spectra of composite materials (m1, m2, and m3)

Figure 4 shows the spectra from subtracting pure cotton from a composite containing chitosan, citric acid, and indomethacin. The non-appearance of peaks in the region from 1680 cm$^{-1}$ to 1750 cm$^{-1}$, indicates the absence of free carboxyl groups in composite materials. A broad band, with increasing intensity, appears at 1570 cm$^{-1}$ depending on the amount of citric acid. The bands in this region are characteristic of the protonated amino groups of chitosan and the carboxylate ion –COO-. The IR band at 1380 cm-1 corresponds to vibrations of the carboxylate ion also. This result indicates that all carboxyl groups of citric acid participate in the electrostatic interaction with chitosan [11, 17]. The other characteristic bands for polysaccharide structure appear at 1150 cm$^{-1}$ due to the C-O-C connection of glycosidic linkage and at 1082 cm$^{-1}$ and 1020 cm$^{-1}$ for C-O stretching skeletal vibrations.

The concentration of citric acid as a chitosan cross-linker on the release of indomethacin in phosphate buffer pH 7.4 at 37 °C for 30 hours was compared. The cumulative indomethacin release from samples m1, m2, and m3 has shown in Figure 5. The initial burst release has been observed due to molecule desorption from the surface for the first 100 minutes. Next, the release has slowed by the more difficult detachment of the molecules inside the chitosan hydrogel. After 24 hours, the released quantity of indomethacin is approximately constant. The citric acid quantity influences the detachment of the drug molecules from composite materials. The released indomethacin increases with the increase of the quantity of the cross-linker. The reason can be the difference in the hydrogel structure and in the drug distribution in the matrix.

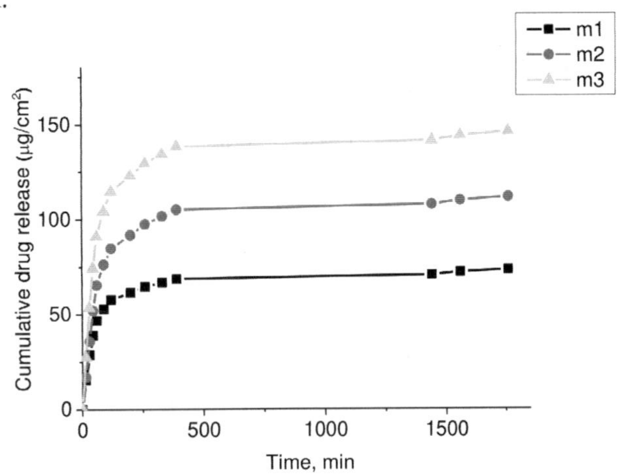


Fig. 5. Release of indomethacin in phosphate buffer pH 7.4 at 37 °C from composite materials (m1, m2, m3) with varying citric acid concentration (at λ =320 nm)

Different kinetic models have been applied to analyze the drug release mechanism from composite materials. Table 2 summarizes the data. The results have shown that the Korsmeyer-Peppas equation describes better the indomethacin release from composite materials.

The obtained correlation coefficient is the highest of the other model equations. Release constant (n) is smaller than 0.89 for samples m1 and m2, and the release mechanism corresponds to the non-Fickian diffusion mechanism. For sample m3 its value is higher than 0.89 and corresponds to super II case transport. In this case, the process is not only diffusion-controlled. The release of the poorly water-soluble drug indomethacin also acquires by swelling and erosion control of chitosan gel.

Table 2. Estimated values obtained by fitting the drug release data to several kinetic models

| | Zero-order | | First-order | | Higuchi | | Korsmeyer-Peppas | | |
|---|---|---|---|---|---|---|---|---|---|
| Sample | $r^2$ | k | $r^2$ | k | $r^2$ | k | $r^2$ | k | n |
| m1 | 0.8817 | 0.3809 | 0.7572 | 0.0108 | 0.9565 | 5.975 | 0.9960 | 0.0302 | 0.8016 |
| m2 | 0.9028 | 0.6223 | 0.7572 | 0.0134 | 0.9685 | 9.7083 | 0.9958 | 0.0182 | 0.8707 |
| m3 | 0.8914 | 0.7903 | 0.7578 | 0.0120 | 0.9623 | 12.368 | 0.9962 | 0.0108 | 0.9856 |

In the agar assay, small zones (about 1.0 - 1.5 mm) were observed around the speciments against both tested strains (Figure 6), which could be explained by slow diffusion due to low hydrophilicity of the agar surface. In liquid medium, the results demonstrated reduction in the cell growth of both strains by all the tested speciments as compared to the negative control. Samples exhibited slightly higher antimicrobial effect against Gram-negative P. aeruginosa than against Gram-positive B. cereus, and m3 sample was found slightly more active than m1 and m2 samples (Figure 7). As can be seen, m3 inhibited about 41% of the growth of P. aeruginosa and about 32% of the growth of B. cereus. Growth reduction of P. aeruginosa by m1 and m2 samples was 35-38% and 27-29% towards B. cereus.

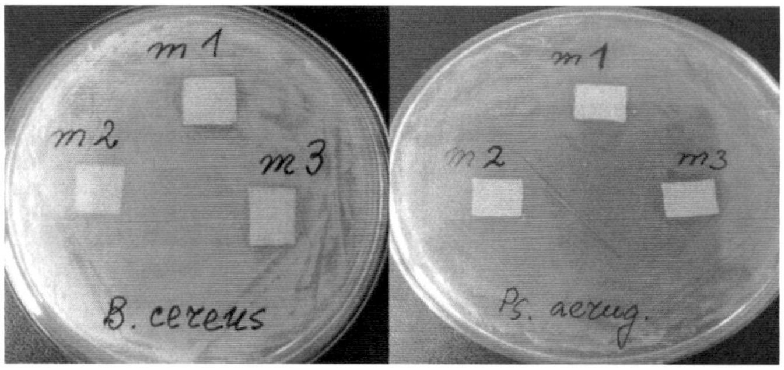


Fig. 6. Agar diffusion test of cotton samples against the model strains B. cereus and P. aeruginosa

It is hypothesized that, the direct contact of bacterial cells with cotton surface contributes to the antimicrobial effect of the treated cotton fabrics. Low water solubility of bioactive compounds does not allow their easy release from the composite materials surface into solution to contact directly with bacterial cells. That explains relatively low growth inhibition of the tested bacteria in solution. The modification of cotton fabric increases its hydrophobicity, thus preventing bacterial growth and bacterial biofilm formation onto the fabric surface.

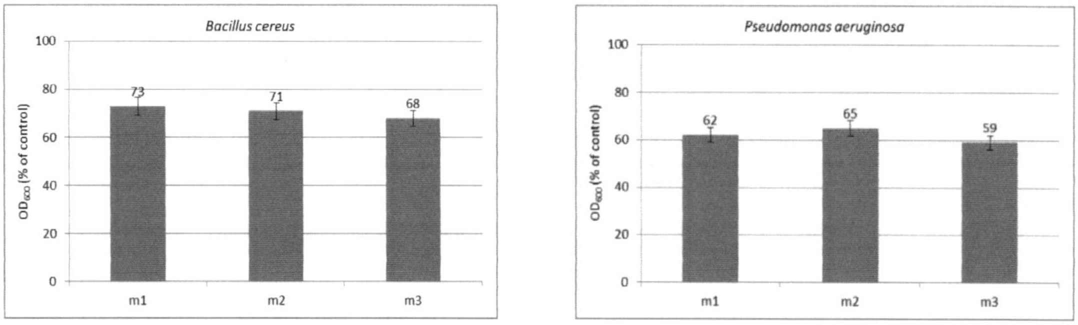


Figure 7. Reduction of the growth of the indicated model cultures by treated cotton fabrics

## 4. Conclusion

New composite materials have been obtained, with potential application as antimicrobial dressings for wounds, releasing an anti-inflammatory biologically active substance (indomethacin). It has been found that as the amount of citric acid used in cotton fabric processing increases, more indomethacin is initially released. After this initial effect, the release of the biologically active substance continues for more than 30 hours in phosphate buffer pH 7.4 at 37 °C. The Korsmeyer-Peppas equation describes the process of the drug diffusion combined with the chitosan gel swelling and erosion. This effect intensifies with the citric acid concentration enhancement. The obtained materials show moderate antimicrobial activity against Gram-positive Bacillus cereus and Gram-negative Pseudomonas aeruginosa bacteria.

**Acknowledgments**
The authors acknowledge the support through Project No КП-06-КОСТ-23 provided by the National Science Fund, Ministry of Education and Science of Bulgaria.

## References

[1] D Atanasova, D Staneva and I Grabchev 2021 *Materials* **14** 930
[2] H Hadjiev, I Rrahnev and P Philippov, In Proceedings of the 38th International Spring Seminar on Electronics Technology (ISSE), Eger, Hungary, 6–10 May 2015; 75–79
[3] S Chatterjee and P C Hui, 2019 *Molecules* **24** 2547
[4] M K Lima-Tenório, E T Tenório-Neto, M R Guilherme, F P Garcia, C V Nakamura, E A G Pineda and A. F. Rubira, 2015 *Chem Eng J* **259** 620–629
[5] M Rani, A Agarwal and Y S Negi, 2010 *Bioresources* **5** 1-43
[6] M C G Pella, M K Lima-Tenorio, E T Tenorio-Neto, M R Guilherme, E C Muniz and A F Rubira 2018 *Carbohydr. Polym.* **196** 233–245
[7] A R Shirvan, M Shakeri and A. Bashari, Recent advances in application of chitosan and its derivatives in functional finishing of textiles, Impact Prospect. Green Chem. Text. Technol. Elsevier 2019 107-133
[8] T Hahn, L Bossog, T Hager, W Wunderlich, R Breier, T Stegmaier and S Zibek, Chitosan Application in Textile Processing and Fabric Coating. Chitin and Chitosan; John Wiley & Sons Ltd., 2019 395– 428
[9] M N Islam, M N Khan, A K Mallik and M M Rahman 2019 *Hazard. Mater.* 379
[10] A B S Agu, P J L Benablo, V ST D Mesias, D P Peneloza JR 2019 *J. Chil. Chem. Soc.* **64(4)** 4610-4612
[11] N Sedyakina, A Kuskov, K Velonia, N Feldman, S Lutsenko and G Avramenko 2020 *Materials* **13**, 1989.
[12] R S Gabbay, R S Kenett, R Scaffaro and A Rubinstein 2018 *Eur J Pharm Sci* **125** 102–109
[13] S Marin, M G Albu Kaya, M V Ghica, C Dinu-Pîrvu, L Popa, D I Udeanu, G Mihai and M Enachescu 2018 *Pharmaceutics* **10** 224
[14] Y Huang, Y Dan, N Dan and Y Chen, 2021 *Polymer Testing* **97** 107129
[15] D Wu, X Chen, T Chen, C Ding, W Wu and J. Li 2015 *Sci. rep.* **5** 11105
[16] F-L Mi, H-W Sung and S-S Shyu 2001 *J Appl Polym Sci* **81** 1700 –1711
[17] J Tanigawa, N Miyoshi and K Sakurai, 2008 *J Appl Polym Sci* **110** 608–615

# Investigation of some factors from the technological processing of leather for upholstery of car seats on their flammability

**M.Koleva[1], D.Zheleva[1], V.Tsanov[2]**

[1]University of Chemical Technology and Metallurgy, Department of Textile and Leather, 8 Kl. Ohridski bul., 1756 Sofia, Bulgaria
[2]Ministry of Interior, MOI Academy, Department of Safety Management and Prevention, Sofia

E-mail: mkoleva2103@abv.bg

**Abstract**. Natural leather is widely used for car upholstery and has a non-flammability requirement. In the present work, bovine leather intended for upholstery for car seats are studied. Various technological schemes of tanning have been applied, including combined; various finishing methods have been applied and two flame retardants have been used, namely *Firex LFS* and *Flacavon B-45*. It was found that the studied methods of combined tanning of investigated leathers do not affect the colour fastness to dry and wet rubbing, but the highest total flame combustion time of the source is given by aluminium-chromium re-tanned leathers and this indicator decreases depending on the type of re-tanning of the tested samples in the following order: chrome, aluminium, vegetable. *Flacavon B-45* showed higher efficiency in terms of flame retardancy of the investigated leathers.

**Keywords:** leather, car seats, flame retardancy, tanning processes

## 1. Introduction

A global trend in recent years is a significant increase in the production of leather for upholstery compared to the leather for clothing and footwear. Numerous indicators that characterize the quality of leather for car seat upholstery include flammability. In our country the current standard refers to protective gloves against thermal hazards, heat and/or flame, as the combustion is performed by burner, the following indicators are determined: flame combustion time and residual smoldering time. A method is known [1], which refers to upholstery of furniture and sources of ignition (smoldering cigarette and gas under pressure, comparable to the effect of lit matches).

The limitation of the flame on leathers depending on the re-tanning and fatliquoring agents has been studied [2]. Standard methods were used: Vertical burning test and oxygen index test [3]. *Caesalpinia Spinosa* fruit extracts have been studied as a pre-tanning agent for the treatment of fire-resistant leathers [4], eliminating the use of aldehydes, synthans and mineral salts in combined vegetable tanning.

IOP Conf. Series: Materials Science and Engineering 1188 (2021) 012005 doi:10.1088/1757-899X/1188/1/012005

Considering the conditions for the possible ignition of the leather during the movement of the vehicle by the driver, the most suitable method is considered for assessment of the ignition process, in which is used as an ignition source smoldering cigarette [5].

In this regard, it is of interest to characterize this method on variants of combined tanning (re-tanning of pre-tanned white leathers) and finished bovine leather for car seats on some performance properties, including and their fire resistance.

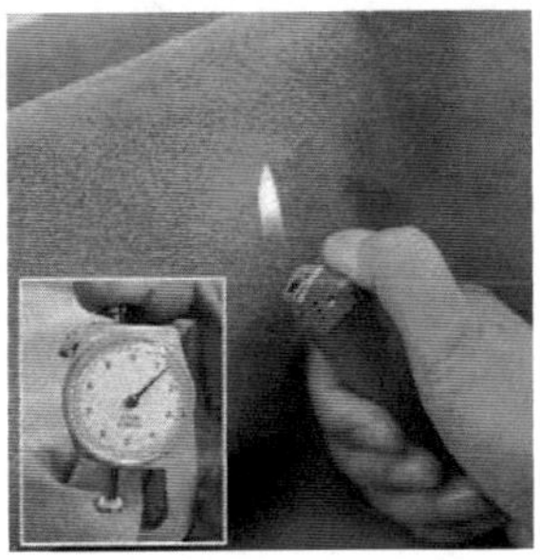

## 2. Experimental

In the present study have the following objectives:

1. Investigation of the method of combined tanning (re-tanning of aluminum pre-tanned bovine leathers) on some performance properties of leather for haberdashery;
2. Study of the influence of the method of combined tanning and processing to increase the fire resistance of bovine leathers intended for car seats;
3. Characteristics of the method: "Furniture: Assessment of the ignitability of upholstered furniture - Part 1: Ignition source: Smouldering cigarette" [1].

To solve the first task, bovine leather samples intended for car seats were processed and the physico-mechanical parameters were determined, as well as the colour fastness of the finish layer to dry and wet rubbing and their fire resistance.

The samples were taken by the method of asymmetrical strips of half bovine leather, treated until the pickling process inclusive.

The method for estimating ignition was applied on 6 parallel samples of treated bovine leathers and a statistical calculation of the results was made.

8 variants wer e studied, 2 samples per variant. The samples from the 1, 2, 3 and 4 variants were pre-tanning by formaldehyde to the wet-white product according to the prospectus data of "Schill & Seilacher" [6].

The samples of the 1 and 2 variants are chrome-retanning, and those of the 3 and 4 - vegetable.

The samples of the 5 and 6 variants are pre-tanning by aluminum, the samples of the 5 variant is chrome re-tanning, and on the 6 sample is vegetable. The leathers from the 7 and 8 variants are vegetable pre-tanning with *mimosa* extract according to a known scheme [7].

After neutralization, dyeing and fatliquoring process of all samples, treatment was performed to impart flammability of samples 1, 3, 6 and 7 with the flame retardants *Firex LFS* (product of condensation of methylene-containing and nitrogen-containing compounds of oxyaryl and inorganic salts), and for samples 2, 4, 5, 8 - with the flame retardants *Flacavon B-45* (combination of halogen and phosphorus salts of organic and inorganic origin) [8]. All samples were finished according to methods for finishing the bovine nappa furniture [9].

According to the given standards, the physico-mechanical parameters [9], the indicators of colour fastness of the finish layer to dry and wet rubbing [10], as well as their flammability according to [5] are controlled.

A statistical calculation of the data for ignition of the bovine leather samples for furniture upholstery was performed according to the indicators: beginning of smouldering /s/ and total flame combustion time of the source (cigarette) /s/.

## 3. Results and discussion

The results obtained for the physico-mechanical properties of aluminium and vegetable pre-tanning bovine leathers for upholstery, depending on the type of their re-tanning, are shown in *Table 1*.

***Table 1.*** *Physico-mechanical results of aluminium and vegetable pre-tanning bovine leathers for upholstery, depending on the type of their re-tanning*

| № variant / Type re-tanning | Elongation at cracking of the face layer of leather, % | Strength of the facial leather layer, N/mm$^2$ | Elongation at leather rupture, % | Tensile strength, N/mm$^2$ DIN EN1336 N/mm$^2$ |
|---|---|---|---|---|
| **Variant 5** Al – Cr - Vegetable | 55,3 | 22,5 | 58,6 | 25,2 |
| **Variant 6** Al - Vegetable - Al | 57,5 | 11,4 | 57,5 | 11,9 |
| **Variants 7, 8** Vegetable - Al | 60,3 | 14,4 | 61,6 | 14,9 |

From the data for the elongation at the cracking of the facial leather layer and the elongation at the leather rupture it can be seen that all the studied variants show relatively high values, the highest being for the vegetable-aluminum tanned leathers.

The data on the strength indicators show that the results for the tensile strength indicator meet the requirements of the standard only for aluminum-chromium-vegetable samples (variant 5), and for the others they are relatively lower. This is probably determined by the chrome tanning process.

*Table 2* shows the results of the durability of the finishing of bovine leather upholstery depending on the tanning scheme, as well as the type of fire protection treatment.

***Table 2.*** *Influence of tanning and flame resistance methods on the resistance to finishing layer of the bovine leather to dry and wet rubbing*

| Tanning scheme / № variant | Dry rubbing resistance, degree | Wet rubbing resistance, degree |
|---|---|---|
| **Variant 1** Aldehyde - Cr | 4/5 | 3 |
| **Variant 2** Aldehyde - Cr | 4/5 | 3/4 |
| **variant 3** Aldehyde - Vegetable | 5 | 3 |
| **Variant 4** Aldehyde - Vegetable | 4/5 | 2 |
| **Variant 5** Al – Cr - Vegetable | 5 | 4/5 |
| **variant 6** Al – Vegetable - Al | 4/5 | 3/4 |
| **Variant 7** Vegetable - Al | 5 | 3/4 |
| **Variant 8** Vegetable - Al | 5 | 2/3 |

The data on colour fastness to wet rubbing for all tested samples meet the requirements of the standard, which for pigmented leather at 250 rpm are at least 3 on the grey scale, except for the samples of variants 4 and 8 (treated with *Flakavon B-45*). *Table 2* shows that the colour fastness to wet rubbing of the tested leathers for variants 4 and 8 does not meet the requirements for the use of *Flakavon*. The treatment of the leathers with this flame retardant leads to a reduction of the degree of resistance to wet rubbing in all variants - 2, 4, 6, 8.

The results for colour fastness to dry and wet rubbing of finished bovine leather upholstery show that they do not depend on the tanning scheme, but to some extent depend on the type of the flame retardant used. This effect can probably be explained by the weaker binding of the flame retardant agent to the dyed and fatliquored leather tissue [12].

The results obtained from the ignition of the tested bovine leather samples, pre-tanned and re-tanned in different variants and treated for incombustibility with both flame retardants, are shown in *Table 3*.

***Table 3.*** *Dependence of ignition of bovine leather samples from the tanning method and from the type of incombustibility treatment*

| Tanning method / № leather sample | | Beginning of flame combustion /s/ | Beginning of smouldering /s/ R = 22s | Total flame combustion time of source, / s / R = 22 s |
|---|---|---|---|---|
| Aldehyde - Cr | **1** | - | 60 | 822 |
| Aldehyde – Cr | **2** | - | 175 | 709 |
| Aldehyde – Vegetable | **3** | 116 | 56 | 542 |
| Aldehyde – Vegetable | **4** | 116 | 238 | 624 |
| Al – Cr – Vegetable | **5** | - | 183 | 592 |
| Al – Vegetable – Al | **6** | 236 | 76 | 692 |
| Vegetable – Al | **7** | 266 | 60 | 606 |
| Vegetable – Al | **8** | - | 124 | 668 |

From the results obtained, the effect of "beginning of flame combustion" can be noted in variants 3 and 4, followed by variants 6 and 7. In most samples, respectively: variants 2, 4, 5, 8, the index of "beginning of smouldering" has higher values, which shows that treatment with *Flacavon* gives better fire resistance to the tested leather samples.

Regarding the indicator "total flame combustion time of the source" has the highest values at the variants 1 and 2 (chrome re-tanned), followed by variants 7 and 8, and lastly: variants 3 and 4 (with combustion effect). These data show that the indicator "total flame combustion time of the source" decreases depending on the type of re-tanning of the investigated samples in following order: chromium, aluminium, vegetable.

From these results it can be concluded that the highest value of the indicator "total flame combustion time of the source" have the leather samples, which mineral re-tanning and those treated with the flame retardant *Flacavon B-45*.

The data from the statistical calculations of the results of 6 parallel measurements for flammability of bovine leather for upholstery, which tanned through combined tanning agents, according to the following indicators: "beginning of smouldering" ($S_{sm}$) and "total flame combustion time of the source" ($C_{total}$) are given *in table 4*.

***Table 4.*** *Flammability indicators statistics of bovine leather for haberdashery*

| **Statistics and indicators** | **Beginning of smouldering /s/** | **Total flame combustion time of source, / s /** |
|---|---|---|
| $R_i$ - arithmetic mean deviation | 192,0 | 704,0 |
| S - standard deviation | 21,0 | 51,0 |
| R - confidence interval /ρ = 0,05; υ = 5/ | 22,0 | 55,0 |
| $S_r$ - random error | 10,9 % | 7,3 % |

The data in *Table 4* show that the random error $S_r$ for both indicators is relatively high, being smaller for the indicator "total flame combustion time of the source" ($C_{total}$) - 7.3%.

These data characterize the studied method of ignition as guaranteeing sufficient uniformity of the respective indicators.

## 4. Conclusions

It was found that the studied methods of combined tanning do not affect the resistance of the finish layer of bovine leather for upholstery towards colour fastness to dry and wet rubbing. It was found that the flame retardant *Flakavon B-45* reduce the resistance of the finish layer of bovine leather for upholstery towards the wet rubbing, but slows the beginning time of smouldering of the tested leathers. The highest "total flame combustion time of the source" show the re-tanning leathers with Al-Cr for haberdashery and those treated with *Flacavon-B-45.*

It was found that the anti-inflammatory treatment of leather for upholstery is more effective when using *Flacavon B-45*.

The tested method for ignition of leather upholstery for car seats guarantees sufficient uniformity of the indicators: beginning of smouldering /s/ and total flame combustion time of source /s/.

## References

[1] EN 1021-1,1993: Furniture - Assessment of the ignitability of upholstered furniture - Part 1: Ignition source: Smouldering cigarette
[2] A. Bacardi, M.D. Borras, J. Sobber, V. Herrrero, J. Jorge, L. Olle, J. Amer. Leather Chem. Ass. (JALSA), 2010, Vol.105, 51-61
[3] Chen Wuyong, Liu Chunxia, Gong Ying, Huang Zan, JSLTC, 2007, vol. 91, 4
[4] L. Olle, C. Casas J. Diaz, S.Sorolla, A. Bacardi, JALSA, 2014, 109(1): 25-31
[5] A. Bacardi, J.Amar, JALSA, 2006, 111 (5), 185
[6] Prospectus data of the Company „Schill & Seilacher"
[7] E. Heidemann, V. Tanigold , VIII Congress the Leather Ind., Budapest 1986
[8] Bo Li, Jiaxun Li, Lixin Li, *et al*, JALSA, 2014, 109(7): 239
[9] Prospectus data of the Company "Langro"
[10] DIN EN 13336: Leather - Upholstery leather characteristics - Guide for selection of leather for furniture
[11] BDS 9050-71: Leather treated, soft. Testing of the coating layer of the leather
[12] B. Duan, Q. Wang, X. Wang, Yun Li, M. Zhang, S. Diao, Results in Physics, 15 (2019), 102717

IOP Conf. Series: Materials Science and Engineering 1188 (2021) 012006 doi:10.1088/1757-899X/1188/1/012006

# Cutting as an Innovative Approach to Surface and Textile Design

**Dafna Stoilkova**

National Academy of Art – Sofia, Faculty of Applied Art,
Textile Art and Design Department
100 G.M.Dimitrov Blvd., 1700 Sofia, Bulgaria

dafna.stoilkova@nha.bg

**Abstract**. We live in an age where new technologies enable materials and techniques, hitherto unconventional for this field, to start being applied in the design of surfaces and textiles. One such technique is cutting. Used for the creation of works with high artistic value, the manual cutting of paper has been developed to the level of an autonomous art form: kirigami – the Japanese art of cutting paper. Inspired by the technique of kirigami, designers and artists today continue to create two- and three-dimensional surfaces and artworks through patterns of cuts in a wide variety of materials such as textiles, polymer foils, wood, metal and others. These cuts can be made by hand, or cut by machine, using technologies such as laser cutting. Novel technologies and materials open the way for an innovative expansion of the characteristics, functions and the aesthetic value of the textiles and surfaces created through these processes. The in-depth study and calculation of the geometry of such patterns of cuts also plays an important role in the scientific research of "smart" surfaces, as well as of surfaces which have the capability to morph from a two-dimensional structure into a three-dimensional one and vice versa. The current research has as its aim to survey how contemporary designers study and develop cutting techniques in unconventional ways, such as creating three-dimensional structures using minimal manipulation, engineering the properties of surfaces and finding alternative ways of constructing objects. The relation between structure, function and aesthetics and the technologies and materials used will be analysed to determine prevalent tendencies in the field.

## 1. Introduction

### *1.1. Cutting as a form of art*

Cutting has existed as a form of art for a long time. Cut images are created on the basis of the principle positive – negative, e.g. material and free space. While other techniques for the creation of pictorial images add extra material, e.g. pigment, to the base material, e.g. paper, in cutting material is removed in order to create an image.

Simplicity and accessibility make the technique of cutting a part of folk traditions of art worldwide. Some of the oldest is the Chinese. The Chinese are thought to have cut out images predating the invention of paper. One of the earliest records of the practice we find in "Records of the Grand Scribe" by Sima Qian, where he describes how a king gave a leaf from a pawlonia tree, cut into the shape of a jade token, to a prince in order to enfeoff him. [1, 2]

This story reveals not only the ancient origins of cutting, but its ritual function. The cutting out of paper images is part of many Chinese folk customs. The images themselves are rich in symbolism, which derives from ancient Chinese philosophy.

Many of the applications of the cutout paper images were decorative, for example as window decoration, or as template for embroidery or the decoration of pottery, etc. A notable example we find in the national dress of the Miao people, who had no writing. Instead, they tell the story of a battle that took place 4500 years ago by means of embroidered patterns, produced with the help of cut templates. [3]

Even before the invention of paper, different sheet materials were being cut: metal sheets, leather, silk. [4] The invention of paper by Cai Lun in 2nd century CE could be considered as the beginning of paper cutting as a form of art. [5] Nowadays the Chinese art of cutting paper is called *jiǎn zhǐ* (Ch. 剪纸 – "cutting paper").

The technology of paper-making was brought to Korea from China along with Buddhism and from there also to Japan in the 6th century CE. [2, 6] One of the ritual applications of cutting was for the manufacturing of *shide* (Jp. 紙垂 – "paper prayers") – objects most often made out of paper through cutting and folding and used as ritual offerings to the deities of the indigenous Japanese religion Shintō. [7] The manipulation of paper in this way creates a structure whose length exceeds the length of the original sheet, simultaneously acquiring a three-dimensional quality as well. Thus cutting goes beyond pictorality – it creates a sculptural effect, an object. During the 11th century CE, *mon-kiri* (Jp. 紋切り – "cutting of crests") which is the art of making cutouts of Japanese family crests out of paper emerges in Japan. [8] What is specific about this practice is the folding and subsequent cutting out, thus fashioning an image with bilateral or radial symmetry. [9] These two cases exemplify cutting *out* and cutting *through* which enable a pattern or structure with repeating elements to be created.

This kind of cutting is nowadays better-known under the term *kirigami* (Jp. 切り紙 – "cutting paper"), coined by Florence Temko in 1962 in her book "Kirigami, the creative art of cutting paper". [10] Today the term is mostly used to connote the cutting out and the cutting through of patterns made of repeating elements and the thus achieved three-dimensional effects. [11] The current article will focus on this aspect of kirigami, its mechanics and application in the sphere of textile and surface design.

### *1.2. The role of materials and technologies for cutting as a technique used for the design of surfaces and textiles*

Initially different sheet materials were used for cutting and later the artform of paper cutting emerged with the invention of paper, which we can observe today being transferred onto other sheet materials, typical for the current age, and even at an invisible, nano-scale. [2] The development of new materials inspires new applications for cutting as a technique for the design of surfaces at any scale – from the nano to macro level.

Contemporary technologies such as laser cutting and others enable the application of cuts in sheet materials such as fabrics, polymer foil, veneer, latex and others, which would otherwise be difficult to manipulate by hand. While in cutting as a form of art the accent lies as much on the pattern being created by the cuts, as on the paper used and virtuosity of the handiwork of the artist, laser cutting is fast, easy and precise and simultaneously enables mass production, which is essential for design.

### *1.3. Various applications of cutting as a technique for the design of surfaces and textiles*

The currently blooming technique of kirigami is a great focus of interest in the fields of design, material engineering, architecture and others, as it offers a simple, economical and effective process for the design of surfaces through structuring achieved by cutting.

In textile design, surfaces with a netted structure are created by the cutting of conventional textiles, leather and so on. Such surfaces have been used in the field of fashion by designers such as Dion Lee

[12], Giorgio Armani [13] and others. Since the beginning of her career, designer Iris van Herpen has been creating structures with the use of cutting. In 2017, in collaboration with architect Philip Beesley, she designed dresses with elastic surface, created by the cutting of a dense pattern of waves in sheets of mylar fabric. The effect achieved is a fast-moving structure with a three-dimensional effect. [14, 15]

Kirigami is a subject of scientific interest, because it enables the direct structuring of particular areas in a sheet of material. Furthermore, the specific qualities of a flat sheet of material are combined with the qualities informed by particular geometries. Research has lead to new forms of *kirigami* such as *nano-kirigami* [2] – kirigami applied at the nano-scale with the aim of informing desired photonic functions – and *kiri-kirigami*, literally "cutting cut paper" [16] – the result of in-depth research on the behaviour of already cut surfaces and their subsequent improvement through additional cuts or notches.

Kirigami-based structures can be "programmed" to shapeshift. Surfaces consisting of such structures could find application in architecture as "smart" windows, which can let in or block sunlight through the opening and closing of the structure [16]; as solar cells, designed to follow the sunlight through morphing their surface structure without the need for moving the entire panel, thus providing maximal energy efficiency [17]; in soft robotics, in the form of a surface that enables crawling like the scales of a snake [18]; or even as shoe soles structured in such a way as to prevent slipping [19], etc.

## 2. Cutting as a tool for the creation of two- and three-dimensional surfaces

The creation of three-dimensional structures using only a two-dimensional sheet material without any addition can be an artistic decision, as well a mathematical, engineering or other solution. The different varieties of kirigami enable design and production processes using minimal intervention and amount of material.

Kirigami uses *cutting through*, *cutting out* and *folding*.

*Cutting through* is the application of a pattern of cuts in a surface, without removing anything from it. Patterns of a decorative as well of a functional character can be cut. From a mathematical point of view, cutting through is similar to division. A mathematical plane (two-dimensional, without volume, consisting only of points) can be theoretically endlessly divided.

*Cutting out* removes parts of a surface, similar to subtraction in mathematics, thus leading to lightening.

Both cutting through and cutting out can create structures which allow for expansion of the surface. The cuts and the cut outs provide the possibility for the opening of extra space, e.g. for the increase in volume of the structure without additions to the surface area.

*Folding* has the reverse function – it tightens, compresses, reduces the volume of a structure, without reducing its surface area.

### *2.1. Using cutting out and cutting through in combination with folding*

Two principal ways for the creation of three-dimensional structures out of a two-dimensional material are folding and cutting. For example, in classical mon-kiri the paper is folded, after which parts of it are cut out, resulting in an image consisting of repeating elements. [8]

Another option is for a pattern of cuts to be applied first, after which the cut sheet to be folded at preliminarily planned places with the aim of the structure acquiring a three-dimensional quality, which is a widely-applied practice in contemporary kirigami. Three-dimensional structures can be created with such techniques. For example, honeycomb-type structures are characterized by lightness and stability, whereby their construction on the basis of a flat sheet can produce functions such as shapeshifting and others. [20]

### *2.2. Plain cutting*

Kirigami does not imply that it is obligatory to cut parts of the material out. The cut motifs can be linear, i.e straight or serrated lines or waves, applied into the material in the form of cuts. When the thus manipulated surface "unfolds" or "opens up", the sheet material is transformed into a three-dimensional structure or even a sculpture. [9]

## 3. Cutting as a tool for informing the qualities of surfaces

All the above-described options allow for the creation of structures with intentionally aimed at qualities such as stability, filigreeness, elasticity, and others, determined by the geometric proportions and the positioning of the folds as well as the cuts. Flexible and adaptive structures can be created, which unfold the advantages of the flat sheet material. Another possibility is to create surfaces which can morph from a two-dimensional into a three-dimensional state and back.

### *3.1. The role of the material*

By applying a pattern of cuts to a flat sheet, its qualities can be altered in such a way that it acquires characteristics which are not intrinsic to it. In this way, the qualities of the material composing the surface remain unaltered, e.g. it is the structuring which changes these characteristics. For example, a piece of paper which is not perceived as elastic can become such by the application of a pattern of cuts. Furthermore, by varying the proportions of the pattern of cuts the degree of elasticity of the sheet of paper can be controlled. [21]

Elasticity is not the only quality which can be obtained by cutting a sheet of paper or another material. This type of structuring enables the flat sheet to form diverse shapes and volumes. A variation of the patterns, used to create elasticity, is also used for the fashioning of honeycomb-type structures, where cutting is combined with folding. [20] Depending on the material used it is possible for the thus formed surface to become a permanent three-dimensional form, e.g. by the use of a metal foil.

### *3.2. Exploring the relation between structure, material and function*

The relation between structure, function and sense perception was the inspiration for my project *Resonance* (supervisor: Prof. Dr. Zane Berzina, Art Academy Weissensee – Berlin, 2015). The project explored the possibility for the creation of elastic structures – nets – out of inelastic polymer sheets using the cutting technique. A pattern was developed through which a sheet could be transformed into an elastic net, which could morph from a two-dimensional structure into a three-dimensional one, and vice versa. This principle is the basis for a collection of adaptive sculptural surfaces, formed out of a variety of sheet materials and manipulated only through strategically positioned laser cuts. [21]

Figure 1. An elastic surface structure obtained using the cutting technique.
A student work by Dafna Stoilkova, supervisor Prof. Dr. Zane Berzina,
Art Academy Weissensee – Berlin, 2015. [21]

## 4. Cutting as an alternative approach for constructing objects

The capabilities of the technique of cutting – the achieving of three-dimensional structures out of two-dimensional surfaces, as well as the alteration and informing of their qualities – lead to the idea that the structuring of surfaces can turn into the constructing of entire three-dimensional objects. The advantage of this alternative method for construction is that it offers the possibility for the fashioning of sculptural forms, using a single two-dimensional sheet of material, thus reducing the number of manufacturing steps.

### *4.1. Constructing zero-waste accessories through cutting*

The technique of cutting allows for the creation of zero-waste products, such as a sustainably produced bag (Project *CUT*, supervisors: Prof. Clara Leskovar, Prof. Doreen Schulz, 2018, Art Academy Weissensee – Berlin). The bag is constructed out of a single piece of sheet material and is manufactured in one single step through laser cutting or a similar technique. [22]

Figure 2. A net bag constructed from a single piece of sheet material using only laser cutting. A student work by Dafna Stoilkova, supervisors Prof. Clara Leskovar, Prof. Doreen Schulz, Art Academy Weissensee – Berlin, 2017. [22]

### *4.2. Cutting as an approach to constructing garments*

A further possibility of the cutting technique is the creation of entire garments, accessories and more. Based on the experience from the projects Resonance [21] and CUT [22], an entire dress is fashioned out of a single piece of material using only cutting as a technique for surface design and construction. The dress turns from a flat cut-through sheet into a three-dimensional functional object when worn on the body.

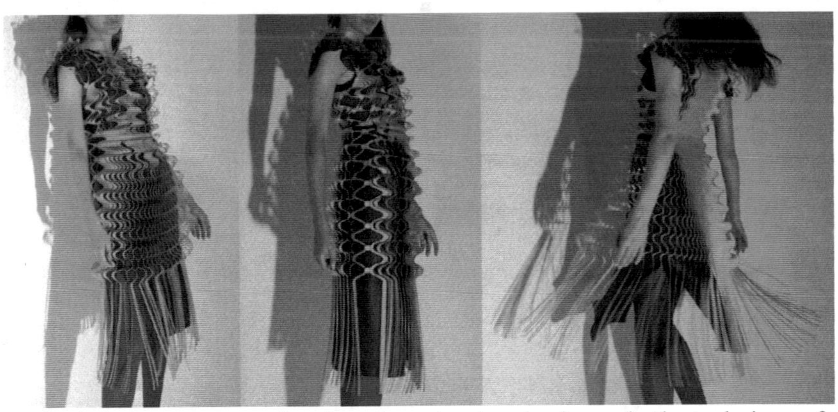

Figure 3. A dress, constructed and surface structured using only the technique of cutting. A work by Dafna Stoilkova, presented during the Global Qipao Invitational Exhibition 2020 at the China National Silk Museum in Hangzhou, China. [23]

## 5. Further aspects of the process of structuring using cutting

The technique of kirigami enriches the palette of ways for the transformation of an idea into material reality, whereby its chief distinguishing quality is that the sole thing required is a drawing which needs to be cut. In this way of structuring, the flat material can attain a two-dimensional pattern or morph into a three-dimensional structure, which informs its functional and aesthetic qualities and can effect a play on perceptions.

### *5.1. Relation between structure, function and aesthetics*

The design process which uses cutting as a tool for the structuring of surfaces can be connected to the solution of a particular mathematical problem. The aimed functionality can be achieved through a pattern with a particular geometry. This specific geometry is by itself a two-dimensional pattern with its own aesthetic qualities. When stretched, the pattern can "come to life", morphing from a drawing into a three-dimensional form. This ability by itself can also be perceived as an aesthetic quality.

Since the mechanism of creating a three-dimensional structure can obviously be found in a two-dimensional pattern, the reverse approach is also possible – by the cutting of an artistic drawing, its pattern can prove to have some unexpected functional features. By virtue of this alternative design process, the quest for artistic expressiveness can turn into a constructive solution. On the other hand, the development of a precisely calculated geometry can result in a structure with particular aesthetic qualities, which could sometimes also be perceived as decorative. Thus, functionality and aesthetics become synchronized by means of using a single technique in one single step – cut.

### *5.2. Relation between structure, material and sense perception*

In the context of this article, surfaces with a three-dimensional quality are formed out of a structure which can be static or kinetic. The structure can be static when its motion is limited across the surface. The structure can become dynamic through a pattern of overlapping motifs, which either repeat or vary, forming nets or similar structures when unfolding. [21] An external activation of such a net by pulling etc., induces movement that causes the three-dimensional structure, for instance, to open or close. In this way the structure becomes "active", kinetic, interactive with its environment. The degree of staticity or dynamicity is determined also by the material used.

While there are structures in areas such as architecture and material engineering created to react to external stimuli such as light, heat, etc., in fashion the mere movement of the human body can create an interaction. This pertains particularly to the experimental dress described above in this article. The morphing from a two-dimensional pattern into a three-dimensional structure and back, kinetically "activated" by the motion of the human body, the confusing materiality of an elastic structure constructed out of an inelastic material – all of these create a play with the senses.

One particular poetic aspect of such structures is that they literally enter a higher dimension – the two-dimensional sheet of material "unfolds" into a three-dimensional structure when kinetically activated by the body. By "doing" so the structure becomes active, alive, which can make its wearer aware of the role of the clothing as an active producer of sensory experiences and thus create an alternative relationship between the wearer, their garment and their environment.

This notion is by no means new. In the beginning of the 20th century Sonia Delaunay created the model Robe Simultanée – "the simultaneous dress". The dress is created on the principle of the "synesthetic melting of time, movement, sound and colour", reflecting Delaunay's concept of oneness of the body and its environment. [24] The Delaunay's simultaneous dress with its combination of colour, composition and material texture invokes sensations which influence the wearer's perception of their own body, their environment and the relation between the two.

## 6. Conclusion

The examples reviewed in the current article have outlined some of the capabilities of the cutting technique, and especially in its kirigami variety: the creation of two- and three-dimensional structures; the altering and informing of qualities of a flat sheet of material; a design process which combines the structuring of surfaces and the construction of objects, among others. All these aspects make kirigami a technique, which will continue to be of interest in various areas of creativity, covering the broad range from the artistic to the scientific and finding conceptual and experimental, as well as practical applications.

**Acknowledgements**

Special thanks to Georgi Krustev for his support with information about objects, created through cutting, and their ritual use.

**All images**



**References**

[1] Nienhauser, WH Jr., editor. The Grand Scribe's Records VOLUME V.l The Hereditary Houses of Pre-Han China, Part I by Ssu-ma Ch'ien. Bloomington: Indiana University Press; 2006. p. 298

[2] Li J, Liu Zh. Focused-ion-beam-based nano-kirigami: from art to photonics. Nanophotonics [Internet]. 2020 Sep 19 [Accessed 2020 Oct 4]; 7(10): 1637–1650. Available from: https://www.degruyter.com/view/journals/nanoph/7/10/article-p1637.xml

[3] Zhilin J. Chinese folk arts / Jin Zhilin, translated by Wang Dehua. Beijing: China Intercontinental Press. 2004.

[4] https://baike.baidu.com/ [Internet] Chinese paper-cut; [Accessed 2020 Oct 4]. Available from: https://baike.baidu.com/item/%E4%B8%AD%E5%9B%BD%E5%89%AA%E7%BA%B8/397259?fr=aladdin (in Chinese, translation obtained using Google Translate)

[5] Encyclopædia Britannica [Internet]. 2020. Paper; [Accessed 2020 Oct 10]. Available from: https://www.britannica.com/biography/Cai-Lun

[6] Robinson N. The Origami Bible. London: Collins and Brown Limited; 2004. p. 10-12

[7] Inoue N. Encyclopedia of Shinto [Internet]. Shide. [cited 2020 Oct 4] Available from: http://k-amc.kokugakuin.ac.jp/DM/detail.do?class_name=col_eos&data_id=22459&fbclid=IwAR17aZ5Z-ZrmLESCki4MvxLd2z5Q323M2JnAFimp4RLFluPz6Ij02_LjmmU

[8] Topping B, Paper Cutting. American Folklife Center/The Library ofCongress Washington, D.C.; 1982 [Internet]; [Accessed 2020 Oct 4]. Available from: https://www.loc.gov/folklife/papercut/Papercut.pdf

[9] Japanology Plus. Japanology Plus - Family Crests [video on the Internet]. 2019 Nov 2 [Accessed 2020 Oct 4]. Available from: https://www.youtube.com/watch?v=U0ltzVp-sI0

[10] Temko F. Origami and Kirigami: 75 Fun-to-Do Projects. New York: Dover Publications, Inc.; 2006. p. 103

[11] Shyu TC, Damasceno PF, Dodd PM, Lamoureux A, Xu L, Shilan M, et al. A kirigami approach to engineering elasticity in nanocomposites through patterned defects. Nature Materials [Internet]. 2015 Jun 22, [Accessed 2020 Oct 4]. p. 1. Available from: http://glotzerlab.engin.umich.edu/home/publications-pdfs/2015/nmat4327.pdf

[12] Singer M, Vogue. Dion Lee Fall 2015 Ready-to-Wear [Internet]. Vogue; 2015 Feb 14 [Accessed 2020 Oct 10]. Available from: https://www.vogue.com/fashion-shows/fall-2015-ready-to-wear/dion-lee

[13] Leitch L, Vogue. Giorgio Armani Spring 2019 Ready-to-Wear [Internet]. Vogue; 2018 Sep 23 [Accessed 2020 Oct 10]. Available from: https://www.vogue.com/fashion-shows/spring-2019-ready-to-wear/giorgio-armani

[14] Iris van Herpen. Between the Lines [Internet]. Amsterdam: Iris van Herpen; 2017 [cited 2020 Oct 10]. Available from: https://www.irisvanherpen.com/haute-couture/between-the-lines
[15] Iris van Herpen. Shift Souls [Internet]. Amsterdam: Iris van Herpen; 2019 [Accessed 2020 Oct 10]. Available from: https://www.irisvanherpen.com/haute-couture/shift-souls
[16] Tang Y., et. al. Programmable Kiri-Kirigami Metamaterials. Advanced Materials. 2016. 29(10):1604262.
[17] Collins G. Kirigami and technology cut a fine figure, together. 2016. PNAS [Internet]. 2016 Jan 12 [Accessed 2020 Oct 11]. Available from: https://www.pnas.org/content/113/2/240
[18] Harvard University. Snakeskin Robot [video on the Internet]. 2018 Feb 21[Accessed 2020 Oct 4]. Available from: https://www.youtube.com/watch?v=SIQma9gyw4E
[19] Young C. Kirigami-Inspired Shoe Bottom Coatings to Slip-Proof Your Shoes. Interesting Engineering [Internet]. 2020 Jun 2 [Accessed 2020 Oct 10]. Available from: https://interestingengineering.com/kirigami-inspired-shoe-bottom-coatings-to-slip-proof-your-shoes
[20] Saito K., Agnese F., Scarba F. A Cellular Kirigami Morphing Wingbox Concept. Journal of Intelligent Material Systems and Structures. 2011. 22:935-944.
[21] Stoilkova D, supervised by Berzina Z, Prof. Dr., Aumann V, Wolf J. Resonanzräume. 2015; [Accessed 2020 Oct 10]. Semester project, Art Academy Weissensee – Berlin. Web: https://kh-berlin.de/projekt-detail/Project/detail/resonanzraeume-1911.html
[22] Stoilkova D, supervised by Leskovar C, Prof., Schulz D, Prof. CUT. 2018; [Accessed 2020 Oct 10]. Meisterschülerin project, Art Academy Weissensee – Berlin. Web: https://kh-berlin.de/projekt-detail/Project/detail/cut-2886.html
[23] China National Silk Museum. Global Qipao Invitational Exhibition 2020 [Internet]. Hangzhou: China National Silk Museum; 2020 [Accessed 2021 Feb 18]. Available from: http://www.chinasilkmuseum.com/zz/info_96.aspx?itemid=27977
[24] Slevin T. Sonia Delaunay's Robe Simultanée: Modernity, Fashion, and Transmediality. Fashion Theory - The Journal of Dress, Body & Culture. 2013 Feb, 17(1):27-54.

# Experimental determination of the frictional characteristics of fabrics made of natural silk

**Umme Kapanyk**

Southwestern University "Neofit Rilski" – Blagoevgrad, Bulgaria

umi12@abv.bg

**Abstract**. Natural silk fabric is directly derived from insect produced cocoons containing large volumes of fibroin protein. The fibroin fibres secreted by the two silk glands of Bombyx mori are smooth and soft and form the structural center of silk. Sericin is a coating with an amorphous structure that bonds the two strands. It is the component hardening the raw thread. The fabric as we know it with low mass per unit area. The level of friction of textile materials depends on a number of test factors - normal load, contact surface area, friction speed, as well as the nature of the textile surface and the direction of friction - warp - warp, warp – weft. In this study, four types of fabric made of 100% natural silk have been examined. They all have different characteristics for mass per unit area, weave, density and linear density of the warp and weft threads. The study makes use of appliance tribometer MXD -02, from Labthink, China to determine the values of static and dynamic friction coefficients. Friction is conducted in different directions and under different pressure. This experimental study has clearly showed that silk fabrics with higher surface mass - display friction coefficient values (at rest and in sliding) greater than 1. Friction index and friction coefficient are influenced by the test direction where friction index in warp direction of one fabric with parallel arrangement of the threads is higher compared to threads in weft direction of another fabric.

## 1. Introduction

A number of technological problems in the sewing industry, incl. layering, cutting, subsequent separation, sewing, packaging, etc. are related to the coefficients of friction of the layers of fabrics in each other.

Next, the user's subjective perception of fabric friction is an important factor, however, touch is not a method for quantifying the friction process. The quantitative parameters of fabric friction, as well as the factors that influence the process are important in the world of textile technologies. The level of friction of textile materials depends on a number of test factors - normal load, contact surface area, friction speed, as well as the nature of the textile surface and the direction of friction - warp - warp, warp - weft.

The surface properties of woven fabrics largely follow their structure [13]. They acquire a plane orientation depending on the mutual contact between the warp and weft threads [11].

Silk fabrics are characterised by low surface mass. This is due to the finesse of the silk threads. The natural silk thread is composed of two proteins - fibroin (about 70-80%) and sericin (20-30%).

IOP Conf. Series: Materials Science and Engineering 1188 (2021) 012007 doi:10.1088/1757-899X/1188/1/012007

The fibroin fibres secreted by the two silk glands of Bombyx mori are smooth and soft and form the structural center of silk. Sericin is a coating with an amorphous structure that bonds the two strands [1]. Sericin is the component hardening the raw thread (Fig. 2).

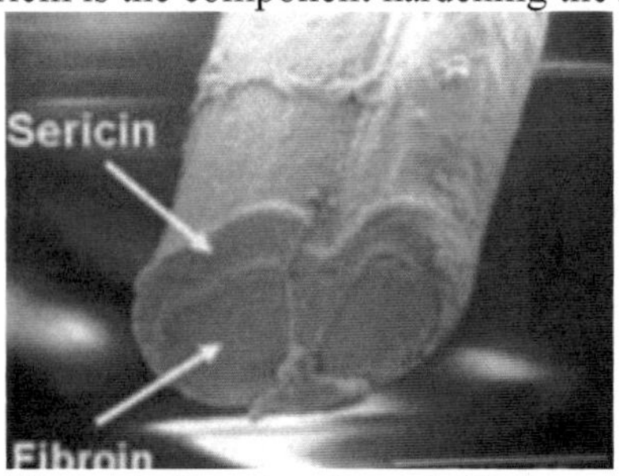

Figure 1. Construction of silk thread [2].

## 2. Experimental part

The research for the present work is performed using four textile materials made of 100% natural silk. The studied textile fabrics have different characteristics. They are produced by company "Svila" AD, Haskovo. Their specific parameters are listed in table. 1 [3].

a) Boyka b) Gergana

c) Veronica d) Neda

Figure 2. Structure of examined fabrics

***Table 1*** Characteristics of examined fabrics made of natural silk

| № | Item | Surface area, g/m$^2$ | Linear density, tex | | Density, н./dm | |
|---|---|---|---|---|---|---|
| | | | warp | weft | warp | weft |
| 1 | Boyka | 76 | N/A* | N/A | 1660 | 350 |
| 2 | Gergana | 80 | N/A | N/A | 1550 | 175 / 175 |
| 3 | Veronica | 170 | N/A | N/A | 2060 | 200 |
| 4 | Neda | 140 | N/A | N/A | 980 | 195 |

Due to the high labour intensity and inaccuracy of the test, the linear densities of the fibers in the finished fabrics have not been determined.

Loss of surface mass and change in thread densities are sufficient characteristics to assess the degree of decoction and shrinkage of the fabrics.

## 3. Measuring devices

Friciton coefficients at rest and at sliding are obtained through work with device MXD-02 from the company Labthink, China (Fig. 1).

The static coefficient is determined by the force measured at the moment when the metal thread connecting force meter and sled straightens and the slide starts sliding on the platform.

The dynamic coefficient is determined by the average value of the measured friction forces. The research was conducted according to BDS EN ISO 8295: 2006. The friction was performed fabric in fabric, the block has a mass of 200 g and the sliding speed is 100 mm / min.

Placing additional weights determines the influence of the pressure force on the friction coefficients of the block. Measurements were performed at pressures of 200 g, 300 g and 400 g.

Friction is performed fabric in fabric so that two layers of the fabric (face side) are placed in different directions. One layer of the examined fabric is attached to the platform so that the direction of warp threads coincides with its direction of movement. The second layer is attached to the mobile sled in a selected direction (warp or weft).

The friction coefficient at rest $\mu 0$ is determined by the force recorded during the test. At this point, the metal thread connecting the force meter and the sled is stretched, after which it begins to slide on the platform. The appliance calculates the average value for the studied friction coefficient during sliding $\mu$, as well as the standard deviations for $\mu 0$ and $\mu$. The numeric values visible on the screen can be printed with the mini printer. The obtained results for $\mu 0$ and $\mu$ are detailed in table. 2.

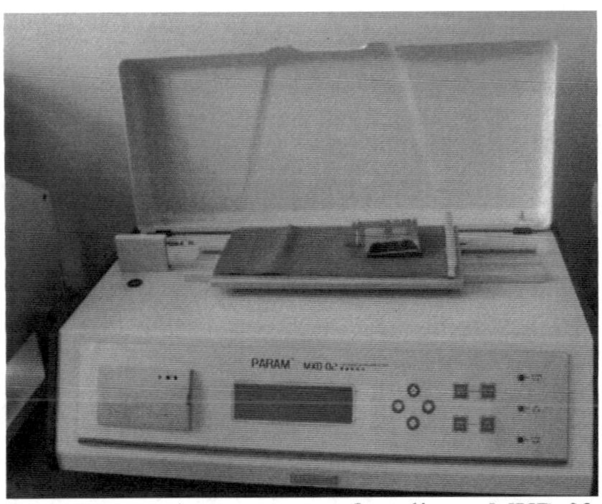


Picture 1. General appearance of appliance MXD-02 [4]

1.Force meter; 2. Movable platform; 3. Lid; 4. Sliding sled; 5. Pressing terminal; 6. Guide rail; 7. Control panel; 8. Display; 9. Mini printer

***Table 2*** Friction coefficient values at rest

| Item | Sled mass with additional load | Friction coefficient at rest with sliding tendency | |
|---|---|---|---|
| | g | WFS-WFS* | WFS-WeFC* |
| Boyka | 200 | 0,333 | 0,410 |
| | 300 | 0,450 | 0,586 |
| | 400 | 0,582 | 0,759 |
| Gergana | 200 | 0,650 | 0,453 |
| | 300 | 0,637 | 0,729 |
| | 400 | 0,837 | 0,883 |
| Veronica | 200 | 1,148 | 1,190 |
| | 300 | 1,438 | 1,486 |
| | 400 | 1,640 | 1,742 |
| Neda | 200 | 1,054 | 1,167 |
| | 300 | 1,241 | 1,434 |
| | 400 | 1,151 | 1,567 |

* Used abbreviations in table. 2: WFS –warp face side; WeFC – weft face side. The aforementioned abbreviations are reused in the tables and graphs in the text to follow. The weight of the test textile piece mounted to the sled is neglected.

***Table 3*** Friction coefficient values at sliding

| Item | Sled mass with additional load | Friction coefficient at sliding | |
|---|---|---|---|
| | g | WFS-WFS | WFS-WeFC |
| Boyka | 200 | 0,292 | 0,360 |
| | 300 | 0,412 | 0,521 |
| | 400 | 0,530 | 0,674 |
| Gergana | 200 | 0,494 | 0,395 |
| | 300 | 0,589 | 0,623 |
| | 400 | 0,774 | 0,773 |
| Veronica | 200 | 0,904 | 0,928 |
| | 300 | 1,184 | 1,211 |
| | 400 | 1,387 | 1,467 |
| Neda | 200 | 0,991 | 0,993 |
| | 300 | 1,279 | 1,284 |
| | 400 | 1,106 | 1,450 |

It should be noted that friction coefficient values normally vary between 0 and 1; however, in some occasion the upper limit of the coefficient is greater than 1. Observing table 2 and 3, friction coefficient (both at rest and at sliding) values greater than 1 can be spotted.

As an example, friction coefficient exceeding the accepted upper limit of 1 is observed for friction at rest with sliding tendency (also known as static friction coefficient) with some metals such as aluminium in aluminium, copper in copper, gold in gold, platinum in platinum etc. Friction force is equal to the normal force when friction coefficient is equal to 1; respectively, the friction force is great to the normal force when friction coefficient is greater than 1. Looking at some existing experimental studies, friction coefficients' aggregate values for some double materials are greater than 1 [5, 6, 7], including materials such as fabrics [8, 9].

One of the main factors found to play a significant role in the friction characteristics determination is the actual contact surface area of friction between fabrics [2]. Increased pressure causes the actual contact area to also increase due to flattening of the threads that occurs.

Experimenting with low sliding speeds (ranging from10 to 500 mm / min.) has resulted in no significant friction force differences according to studied experimental literature [11, 12]. The sliding speed used throughout this study is 100mm/min. That speed is deemed acceptable due to the above mentioned fact (friction speed influencing fabric layers relative to each other in insignificant).

The following logarithmic expression (1) determines the relationship between normal load and friction force [2, 8, 9, 10]:

$$log\left(\frac{F_i}{B}\right) = logC + n.log\left(\frac{N_i}{B}\right) \quad \text{or} \quad \frac{F_i}{B} = C.\left(\frac{N_i}{B}\right)^n \tag{1}$$

i = 1, 2,... m; B - contact area (measured in m2); C - friction parameter (measured in Pa1-n); n - friction index (without dimension); N - normal pressure (in N); F - friction force (in N); m - number of experimental observations.

For cotton and polyester /P/PE/ mixed textile materials with different ratios, experimental studies find that the logarithmic dependence approaches a linear one [4, 5, 6].

Three tests are performed for each studied fabric; an arithmetic mean value is then determined for each one. Different directions and applying different pressure is how log (Fi / B) and log (Ni / B) are determined. The pressure applied can be changed by loading the sled with an additional weight. These weights are added to the sled's own weight of 200g. The weight of the mounted to the sled piece of fabric is neglected with each test and for all studied fabrics as that weight is insignificant and does not alter the results. The room temperature and humidity conditions during all tests are kept to 22°C and 70% respectively.

Log (Fi / B) and log (Ni / B) are calculated after friction factor and friction index are determined and normal compressive forces Ni and friction forces Fi are calculated.

The below equation of linear regression is formulated:

$$y=a+x.b \tag{2}$$
$$x= lg(N_i/B)\ ; \quad y= lg(F_i/B); \quad a=lgC\ ; \quad b=n$$

Friction factor R (also known as "correlation friction coefficient" or "composite friction coefficient") is calculated based on the following dependence [3]. The calculation also makes use of friction index and friction parameter.

$$R = \frac{C}{n} \tag{3}$$

Table 4 presents the results for friction index for friction at rest and at sliding, friction parameter, and friction factor.

***Table 4.*** Values of frictional characteristics at rest and sliding

| Item | Direction of the fabric | Frictional characteristics at rest | | | Frictional characteristics at sliding | | |
|---|---|---|---|---|---|---|---|
| | | Index | Parameter | Factor | Index | Parameter | Factor |
| | | $n$ | $C$ | $R$ | $n$ | $C$ | $R$ |
| | | - | $Pa^{1-n}$ | $Pa^{1-n}$ | - | $Pa^{1-n}$ | $Pa^{1-n}$ |
| Boyka | Warp FS-Warp FS | 0,570 | 0,352 | 0,617 | 0,438 | 0,010 | 0,017 |
| | Warp FS -Weft FS | 0,380 | 0,013 | 0,034 | 0,430 | 0,011 | 0,025 |

| | | | | | | | |
|---|---|---|---|---|---|---|---|
| Gergana | Warp FS-Warp FS | 0,387 | 0,144 | 0,373 | 0,060 | 0,041 | 0,110 |
| | Warp FS-Weft FS | 0,413 | 0,010 | 0,025 | 0,452 | 0,009 | 0,020 |
| Veronica | Warp FS-Warp FS | 0,555 | 0,154 | 0,277 | 0,336 | 0,081 | 0,292 |
| | Warp FS -Weft FS | 0,476 | 0,140 | 0,294 | 0,301 | 0,071 | 0,235 |
| Neda | Warp FS-Warp FS | 0,890 | 0,618 | 0,694 | 0,160 | 0,039 | 0,056 |
| | Warp FS -Weft FS | 0,610 | 0,220 | 0,361 | 0,399 | 0,077 | 0,226 |

## 4. Results analysis

This experimental study has determined that silk fabrics with higher surface mass - Veronica and Neda, display friction coefficient values at rest and in sliding greater than 1. Friction coefficient value equal to 1 means that the friction force is equal to normal force; when friction coefficient values are greater than 1 then the friction force is greater than normal force.

It was also determined that the friction factor for friction at rest is higher than that for sliding.

The friction characteristics - friction index, friction parameter and friction factor (with minor exceptions) have higher values in the direction of warp – warp.

## References

[1] Neznakomova M., Textbook of Textile Materials Science, Technical University of Sofia, 2010. Рахнев И. Р. (2006) Метрологичен подбор на сплитка за работно облекло, сп. Текстил и Облекло, ISSN 1310-912X, брой 8/2006, стр. 21-27, София, 2006.

[2] https://cosetex.it/wp-content/uploads/2017/05/Silk_Fibroin_Sericin_Amino_acids.pdf

[3] Diana Germanova-Krasteva, Rayka Chingova, Sasho Aleksandrov, Umme Kapanyk Study of the effect of finishing treatment on the coefficients of friction of fabrics from natural silk Sp. Textiles and clothing, no. 10.2019 p. 318 / extended summary in English/.

[4] Chingova, R., Germanova-Krasteva D. Experimental study of the frictional characteristics of cotton fabrics with twill 3/1, Mechanics of Machines, Book 1, 2019, ISSN 0861-9727, pp. 27-31.

[5] Chingova, R., Study of friction in some tissues, Monograph, University Publishing House "N. Rilski "- Blagoevgrad, 2013.

[6] Leach, S. Coefficient of Static Friction: A Simple Experiment for Applied Mechanics Students. Proceedings of the National Educators 'Workshop: Update 2004, Phoenix, AZ, 2004.

[7] Hirani, H., Fundamentals of Engineering Tribology with Applications, 1st Edition, Kindle Edition, 2016.

[8] Apurba Das, V. K. Kothari and N. Vandana, A Study on Frictional Characteristics of Woven Fabrics, AUTEX Research Journal, Vol. 5, No3, September 2005, pp 135.

[9] Sülar, V, E. Öner, A. Okur, Roughness and frictional properties of cotton and polyester woven fabrics. Indian journal of Fiber end Textile Research, vol. 38. December 2013, pp 349 – 356.

[10] Chingova, R.,(2018), Experimental study on the friction characteristics of cotton fabrics before and after finishing, Knowledge, Vol. 26.5. pp.1513-1520, Skopje, 2018.

[11] Rahnev Ivelin and Rimini Gaetano, Isotropy equilibrium of the double woven fabric with cotton face and wool reverse fibrous compositions, Proceedings, Corfu, Greece, 2017. IOP Conference Series: MSE, Volume 254, 2017, 17th World Textile Conference AUTEX 2017, Accepted papers received: 6 October 2017, Published online: 8 November 2017, DOI: 10.1088/1757-899X/254/9/092003EID: 2-s2.0-85035002762.

[12] Bhuvana, D., Raghunathan, Subramaniam. Studies on frictional behaviour of chitosancoated fabrics. Autex Research Journal, vol. 6, no. 4, December 2006.

[13] Rahnev I. (2006) Work cloth weave metrological selection, Tekstil i Obleklo, 2006, 2006(8), ISSN 1310-912X, issue 8/2006, pp. 21-27, Sofia, 2006.

# Relaxation function experimental study of the warp strain at slow and regular weaving motion

**I Rahnev**

Quality Control Department of E. Miroglio EAD, Sliven, Bulgaria

Ivelin.Rahnev@emiroglio.com

**Abstract.** The combination of the nonlinear rheological models to the simple pendulum in the conditions of the forced oscillations simulates the action of the let-off motion with weights. The experimental study considers the dynamic balance during the unfolding of the warp in function of the adjustments of the let-off motion. The stress of the warp threads carries out the mechanical transition from the loom's adjustments towards the fabric quality. A rotatable plan of factorial experimentation indirectly explores the process of interaction of threads during the intersection.

## 1. Introduction

In order to arrive at practically applicable solutions, according to tradition we separately consider the quality of the yarns, the weave and the weaving technology as the main factors conditioning the general quality of the fabric. Technological management or optimization of the weaving cycle provides for the variation of various machine parameters, the complex influence of which determines the interaction between the warp yarns and the picks. At this initial research stage, we consider the behaviour of the yarns at the time of levelling the frames because of warp pulling and weft density.

During the weaving cycle, the textile threads are subjected to the action of the working bodies of the loom at the expense of the driving force. The technological mode is based on the tense elastic state of the main threads and the main weave regulator simultaneously controls and regulates the set parameters. Unlike other organs of the loom, this is the only mechanism that interacts and whose action depends on the current state of the processed raw material.

The objective of this article is the mechanical characteristic of the fabric as a function of the warp tension. Also, this is an operation of a let-off weave regulator with weights. The aim is to establish the reactions of the mechanism at different settings. The methodological goal is directed towards determining the limits of machine settings and their functional control through the stages of optimization. The tasks performed include the description of the experiment, the realization of the experimental plan, the laboratory examinations and the statistical processing of the data, the composition of the experimental models and the analysis of the results.

## 2. Research

The tests are carried out on the weaving machine "MAV-S-RPC-185". The simplicity of the mechanisms of the longitudinal movement of the warp threads eliminates parasitic effects, figure 1.

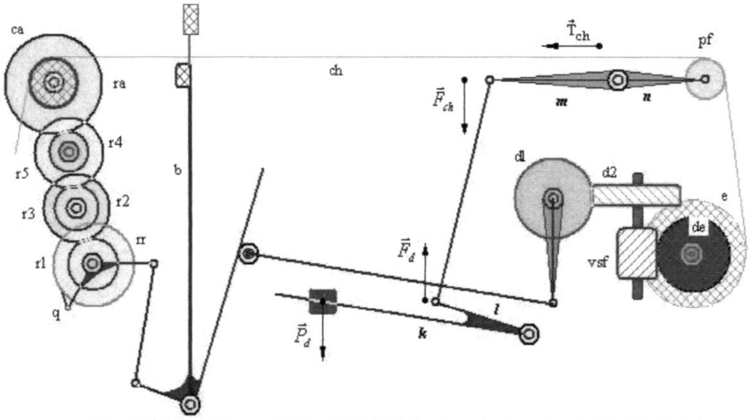


Fig.1, Dispositif du mouvement longitudinal des fils de chaine, système "MAV"

Figure 1, Let-off and take-up motions, principles

The combination of negative let-off motion with positive pick-up regulator is the most widely used technological aggregate for simultaneous control and protection of warp yarns. The two mechanisms are independent on the drive and mechanical settings. As the main organs of the loom, they operate jointly with the warp threads. The functional alliance is expressed in relentless support of constant tension to meet technological demands.

The technological impact of the weaving machine, achieved by warp unwinding and picking, defines the stress state of warp yarns. For one article reviewed, technological optimization consists in modifying controllable impacts, such as machine settings.

The "MAV" system let-off motion is characterized by diversity and stability of the settings, the only spring of which is the chain. This system of levers functions as a balance that maintains the technological regime by balancing the warp tension applied to the thread carrier and the loaded weights. We match the let-off scale to constant weights by the formula:

$$T_{ch} = \left(\frac{P_d.k}{n}\right).\left(\frac{m}{l}\right) \Leftarrow \left[\vec{F}_d + \vec{F}_{ch} = \vec{0}\right] \Leftarrow \begin{cases} P_d = const; n = const; k = const \\ m = tab(1,...,6,...,11cm) \\ l = tab(2,4,6cm) \end{cases} \quad (1)$$

The functioning of the let-off regulator with weights consists in the opposition and in the balancing of the spring force of the tensioned warp threads with the constant gravitational force of the applied mass, fig. 1. The combination of a lever system and an elastic connection receives a cyclic load from the driving machine bodies. The operation of the mechanism is considered by dividing it into three working groups: drive, rheological element and lever system.

The periodic drive of the working bodies in the weaving machine follows the principle of fabric formation and represents the transformation of the rotational movement into reciprocating movements [1]. The warp threads receive the linear motion, which is characterized by the angular velocity of the main shaft and the vertical stroke of the threaded system. As a result, the warp threads periodically receive relative elongation -$\{\varepsilon_\omega\}$, with amplitude and frequency according to the design and speed of the machine:

$$\varepsilon_\varpi = \varepsilon(f_0).\sin \omega.t \quad (2)$$

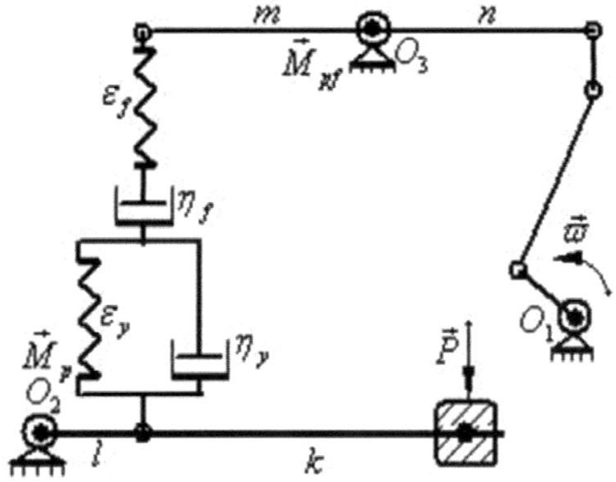


Figure 2, let-off motion cinematic principle

Rheological model: general character of the nonlinear longitudinal deformations of the textile threads is modelled by the analogy series of elastic springs and viscoelastic shock absorbers [2]. The peculiarities of linear textiles require the sequential connection of Maxwell's fluid, which symbolizes the behaviour of twisted fibres: $\{\varepsilon_f, \tau_f\}$, with the solid body of Kelvin -Voight, representing the behaviour of the warp threads multitude: $\{\varepsilon_y, \tau_y\}$.

The stresses and strains of the threads are distributed in the yarns and fibres according to their properties and together form the reaction of the warp:

$$\begin{vmatrix} \varepsilon_{\omega} = \varepsilon_{f} + \varepsilon_{y} \\ \tau_{\omega} = \tau_{f} + \tau_{y} \end{vmatrix} \quad (3)$$

The general viscoelastic behaviour of the base is explained by the different distribution of the external load in the textile structures. While the stress $\tau_f$ applied to the fibrous analogue is distributed uniformly, the final elongation $\varepsilon_f$ of the rheological model of the thread is the sum of the elongations of the elastic and viscoelastic elements:

$$\frac{d\varepsilon_f}{dt} = \frac{1}{E_f} \cdot \frac{d\tau_f}{dt} + \frac{1}{\eta_f} . \tau_f \begin{cases} \varepsilon_f = \varepsilon_f^{\varepsilon} + \varepsilon_f^{\eta} \\ \tau_f = \tau_f^{\varepsilon} = \tau_f^{\eta} \end{cases} \quad (4)$$

The set of main threads reacts back by means of the same total elongation $\varepsilon_y$ and total stress $\tau_y$:

$$\tau_y = \eta_y . \frac{d\varepsilon_y}{dt} + E_y . \varepsilon_y \begin{cases} \varepsilon_y = \varepsilon_y^{\varepsilon} = \varepsilon_y^{\eta} \\ \tau_y = \tau_y^{\varepsilon} + \tau_y^{\eta} \end{cases} \quad (5)$$

Differential equations (4) and (5) of the rheological model reflect the simultaneous influence of the elastic and reducing components in the elongation of the warp.
Lever system: transmission circuit of the regulator is reduced to two groups of levers located on both sides of the observed rheological analogue. The motor impulse passes through the whole mechanism and opposes the torque generated by the concentrated mass. The lever with the applied mass, thus connected to the rheological element, builds the model of a pendulum subjected to forced oscillations.
The dynamic equilibrium of the completely mechanical system is found in the equalization of the dynamic moments of gravity and of the rheological element applied to the loaded lever:

$$\left.\begin{matrix} \vec{P} = m.\vec{g} \\ \vec{M}_p = \vec{P} \wedge k.\vec{e}_k \end{matrix}\right\} \{D_{O_2}^k\} = \{D_{O_2}^L\} \begin{cases} \vec{F}_{\omega} = S_y . \vec{\tau}_{\omega} \\ \vec{M}_{\omega}^{l} = \vec{F}_{\omega} \wedge l.\vec{e}_k \end{cases} \quad (6)$$

The level of gravity, or the angle of the lever with the horizontal, fulfils the basic equation of forced oscillations:

$$\frac{d^2h}{dt^2} + \frac{\eta_{\omega}}{D_{O_2}\{m\}} . \frac{dh}{dt} + \frac{E_{\omega}}{D_{O_2}\{m\}} = f_0(\varepsilon_{\omega}).\sin \omega t \quad (7)$$

Evaluation of the synthetic model: presence of models with mathematical description direct the forecast for the behaviour of the studied mechanical system mainly to the influence of the possible machine settings and neglect of the constant parameters. As a result, it can be expected that the increased dynamic torque will lower the average level of the lever, as well as reduce the amplitude of its oscillations.
Experimental work was performed on a weaving machine "MAV-S-RPC-185". The main regulator has a negative drive and a dependent action with respect to the set machine mode.
The integral force of the tensioned base creates a torque around the axis of the cross, whose unauthorized rotation and unscrewing of the base is prevented by the gear: worm - gear. The control piston with the roller performs a reciprocating motion under the action of the propeller and periodically drives the worm.
The position H of the control roller relative to the axis of the piston determines the stroke of the piston and finally - the rotation of the main cross. The control roller is the actuator of the automatic adjustment, and the occupied moment height h reflects the control signal obtained from the dynamic equilibrium of the mechanism.

Automatic control: warp threads rest on the traction cross cylinder pf and by means of the generalized force $\vec{T}_{ch}$ of their tensioned state they create a torque $\vec{M}_{pf}$ around the axis $O_3$ of the traction cross lever. Changes in the stresses of the main threads change the torque balance in the regulator and cause the guide lever to rotate and move the current level of the control roller.
Machine settings: transmission function of the lever system depends on the position of the connecting rod on the free arm m of the traction lever and on the adjusting arm $l$ of the guide lever. The ratio $m/l$ connects the mass and rheological torques and its value determines the average level of static balance between the tensioned warp $\vec{T}_{ch}$ and the applied weight $\vec{P}$. The size of the masses $\sum m_i$ and their location k on the guide lever determine the torque $\vec{M}_p$ of the symbolic pendulum. The change in the machine setting of the static balance between the force of gravity and the tension of the base is determined by the design parameters of the regulator:

$$T_{ch} = P.\frac{k.m}{n.l} \Leftarrow \left[\vec{F}_d + \vec{F}_{ch} = \vec{0}\right] \Leftarrow \begin{cases} P_d = \sum m_i \\ m = tab(5 \div 16cm) \\ l = tab(5,7,9cm) \end{cases} \tag{8}$$

Observable parameters: the position of the guide lever is the main metric factor, which is the result of mechanical control and automatically adjusts the regulator by the height of the guide roller. This combines the instantaneous result of dynamic equilibrium and the regulatory response of the mechanism.
Conducting experiments: in the first stage, we studied the static equilibrium of the mechanism under conditions of minimum angular velocity of the main shaft of the loom. In the period of one weaving cycle, we currently measured the distances between the cotton axis and the control roller. The obtained values in tabular form make up the diagram of the level of the roll depending on the interaction between the working bodies of the loom and the textile threads, fig. 3.

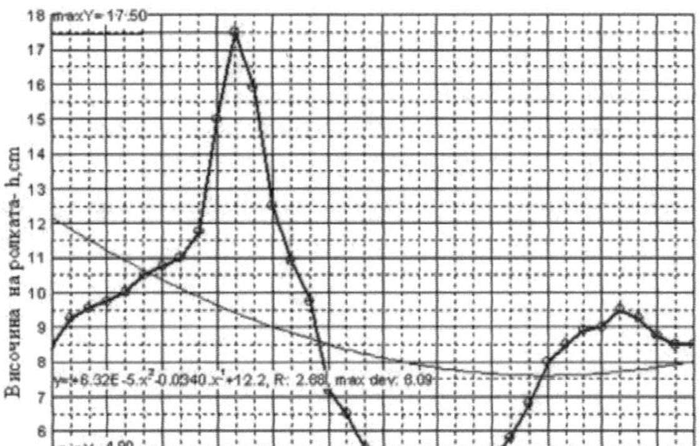


Figure 3, Static equilibrium diagram

We measured the running tension of the warp threads by a mechanical "Deuta-Werke" tensiometer between the beam and the thread carrier when closing the shed of warp threads. The weaving machine is warp loaded with cotton type whose yarns are twisted ring spun yarns and the weft is OE yarn of cotton, Table 1.

Tableau 1, Caractéristiques des fils

| Valeurs nominales | Caractéristiques de laboratoire | | | | | | | |
|---|---|---|---|---|---|---|---|---|
| | Masse linéique, tex | | Torsion, $m^{-1}$ | | Force maximale,cN | | Allongement, % | |
| Tt + composition | Tt | $\sigma$ | Tk | $\sigma$ | max $F$ | $\sigma$ | $\varepsilon$(max $F$) | $\sigma$ |
| 26x2, 33/67-Vi/Pest | 53.84 | 0.726 | 316.7 | 18.97 | 1165.5 | 59.573 | 15.42 | 0.5807 |
| 28x1, 100-Coton | 29.20 | 0.405 | 356.5 | 51.98 | 315.52 | 33.973 | 5.14 | 0.6167 |

The weave pattern is canvas/plain weave for the entire experience.

Experimentation plan: Functional independence and convenient access to settings allow controlled variation of factors in the spherical exploration domain. The generalized parabolic shape of the rheological behaviour of textile products requires modelling at the 2nd degree response surface. These conditions oriented the present research towards the factorial experimental design at isovariance by rotation whose polynomial model for the k-th mechanical property acquires the form:

$$\hat{Y}_k(X_1, X_2) = \beta_0 + \beta_1.X_1 + \beta_2.X_2 + \beta_{12}.X_1.X_2 + \beta_{11}.X_1^2 + \beta_{22}.X_2^2 \quad (9)$$

The preliminary tests revealed the practical limits of the experimental field of interest. The calculated values of the natural variables make up the Experiment Matrix, Table 2.

Tableau 2, Plan d'expérimentation

| Facteurs | Niveaux des facteurs | | | | | | | | | | | | | | | | | |
|---|---|---|---|---|---|---|---|---|---|---|---|---|---|---|---|---|---|---|
| | 1 | | 2 | | 3 | | 4 | | 5 | | 6 | | 7 | | 8 | | 9 | |
| Duitages-Dd, d/cm | - | 16 | + | 18 | - | 16 | + | 18 | α | 19 | -α | 15 | 0 | 17 | 0 | 17 | 0 | 17 |
| Balance, ***m/l*** | - | *11/4* | - | *11/4* | + | *4/4* | + | *4/4* | 0 | *6/4* | 0 | *6/4* | α | *11/2* | -α | *1/6* | 0 | *6/4* |

## 3. Results

Running the experiment: technological conditions of the study base ensured the practical realization of the planned experiment. Randomization of variants and accurate reproduction of planned parameters are part of the customary procedures that we followed in adhering to standard requirements for sample preparation. In the middle of each variant, we measured the tension of the warp threads at different places across the width of the warp and of the threads threaded in different frames.

Following the main theme, the research considers the behaviour of the fabric during tensile loading in warp and weft directions. As a result, we organized, apart from two couples, qualitative indices describing the mechanical behaviour by the maximum tensile force and the corresponding elongation.

Composition of the experimental models: Statistical processing of laboratory data yielded estimates of the properties examined. The nearly equal coefficients of variation of the mathematical estimates referred to them as the natural reactions of the planned factor levels, Table 3.

Tableau 3, Valeurs des réactions expérimentales $\hat{Y}_k(X_1, X_2)$

| Niveaux de facteurs | | *1* | 2 | *3* | *4* | *5* | *6* | *7* | *8* | *9* |
|---|---|---|---|---|---|---|---|---|---|---|
| Tension de chaîne : Tch, cN | | *82.32* | *138.71* | *157.55* | *253.94* | *200.89* | *92.86* | *236.71* | *102.04* | *138.0* |
| Ch | Force - maxF, N | *1150.7* | *1116.6* | *1147.3* | *1198.2* | *1124.4* | *1098.3* | *1176.8* | *1112.4* | *1133.1* |
| | Allongement, ε (maxF), % | *22.928* | *31.785* | *28.822* | *30.375* | *32.495* | *28.693* | *26.099* | *29.378* | *26.970* |
| Tr. | Force - maxF, N | *238.24* | *286.67* | *244.57* | *312.96* | *313.17* | *220.70* | *268.94* | *234.97* | *270.03* |
| | Allongement, ε (maxF), % | *8.375* | *9.294* | *8.690* | *10.015* | *9.768* | *9.121* | *9.355* | *8.165* | *8.609* |

Based on the experimental results and by the known algorithm of regression analysis we calculated the coefficients of the polynomial models. The response surface equations passed the alternative control of dispersion analysis as against the meaning of the coefficients, as well as for the veracity of the models. Preliminary tests correctly determined the experimental centre and area of interest and we obtained models adequate for the maximum number of significant regression coefficients, Table 4.

Tableau 4, Coefficients des modèles polynomiaux

| Coefficients de modèles | | $b_0$ | $b_1$ | $b_2$ | $b_{12}$ | $b_{11}$ | $b_{22}$ | $v$ | $F_r$ |
|---|---|---|---|---|---|---|---|---|---|
| Tension de chaîne : Y1 (Xi) | | 140.88 | 38.195 | 47.614 | 10.00 | 2.999 | 14.249 | 7 | 0.5714 |
| Ch | Force – Y2(Xi) | 1131.2 | 17.56 | 39.55 | 24.75 | 9.45 | 22.45 | 7 | 5.35 |
| | Allongement, Y3 (Xi) | 26.961 | 1.958 | -0.533 | -1.356 | 1.911 | -.0131* | 8 | 3.727 |
| Tr | Force – Y4(Xi) | 270.974 | 34.517 | 10.351 | 5.55 | 1.804* | -8.679* | 8 | 6.029 |
| | Allongement Y5 (Xi) | 8.622 | 0.412 | 0.366 | 0.109 | 0.414 | .0659* | 8 | 5.781 |

Interpretation of models: by having validated models, we can access the graphic interpretation of the results. The equations, intrinsic to bi-factorial planning, plot three-dimensional response surfaces. This visualization shows the characteristic trends in the functional variation of the observed properties in dependence on the factorial variables. The use of existing graphics applications presents in understandable form the experience performed.

Discussion: state of stress of the warp threads reflects in almost linear form the technological parameters of the weaving cycle, figure 4. With the enlargement of the picks comes the increase in tension, so the increased draft of the let-off motion scale increases the tension. The influence of the let-off motion predominates the variation of the picks. The minimum voltage corresponds to the minimum factorial levels and the maximum voltage is a result of the simultaneous maximum levels of the factors. The centre of the experimental area defines the mean values of the chain tension. The maximum tensile force (the force maxF) per chain increases in the paraboloid area of response to clearly manifested factor interactions, figure 5.

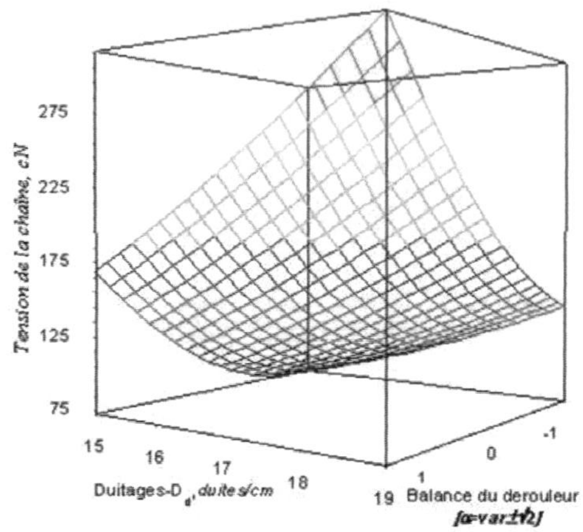


*Figure 4, Warp stress state*

**Fig.2, L'état de contrainte de la chaîne**

Under the conditions of minimum weights, the increased draft leads to a sharp increase in strength. At maximum edits, we observe a parabola symmetrical at least in the centre of the factor: let-off motion balance. The let-off pull opposes the weft insertion. We observe a decrease in force at increased debit and the increase in force is at reduced debit. The minimum value of the force is at the maximum weft density and the maximum debit. The combination of the maximum levels of the factors gives the maximum value of the maximum force. The existing middle zone occupies the diagonal in the experimental domain.

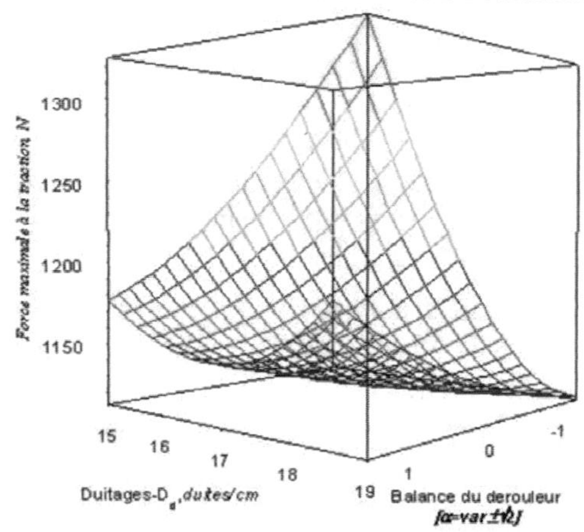


*Figure 5, Warp direction breaking force*

**Fig.3, La force à la traction par chaîne**

The maximum tensile force per weft is characterized by the linear increase as a function of the draft and by parabolic evolution according to the pickings, figure 6.
At a minimum, the variation of the draft leads to the parabolic reaction of the force with a maximum at the centre. At the maximum output the force grows harmoniously up to the common maximum.

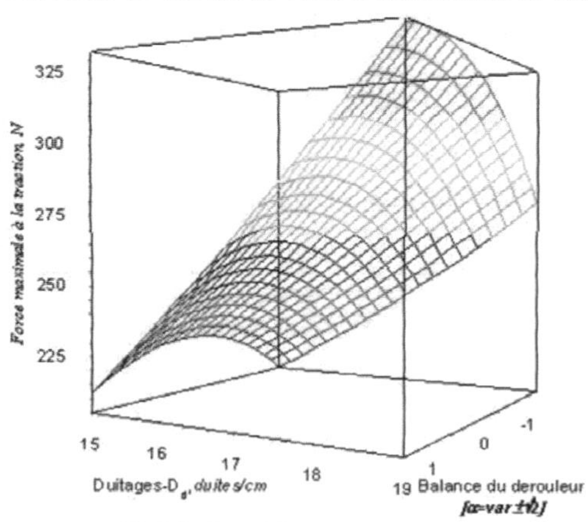


*Figure 6, Weft direction breaking force*

**Fig.4, La force à la traction par trame**

As the weft density increases, the strength increases and the draft increases the slope. The minimum factorial levels define the minimum value of the force and the maximum levels - the maximum value.

The percentage elongation at maximum force (the elongation ε (maxF)) per chain describes the response surface to the concave character. The minimum zone and the single maximum are clearly illustrated there, figure 7.

Under small weft conditions, draft increases elongation, while at high weft density, high elongation is a result of increased debit. The elongation according to the dispensing depends mainly on the action of the let-off motion. At maximum draft, the aspect ratio is parabola with a minimum in the centre. The minimum draft conditions the abrupt increase in elongation with increasing weft density.

*Figure 7, Breaking elongation by warp*

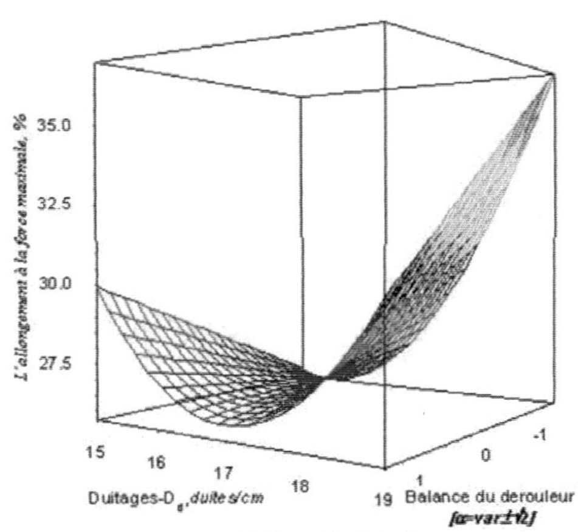

Fig.5, L'allongement par chaine

The warp elongation response surface represents asymmetric paraboloid at the minimally manifested area, Figure 8.
The reduction in warp debit results in a harmonious increase in elongation without considerable influence on the weft density. The variation of the weights is related to the parabolic evolution of the elongation, the minimum of which occupies the central zone. The minimum elongation is a result of the minimum draft and the "average weft density. The maximum elongation reflects the combination of the maximum factor levels.
Figure 8, Breaking elongation by weft

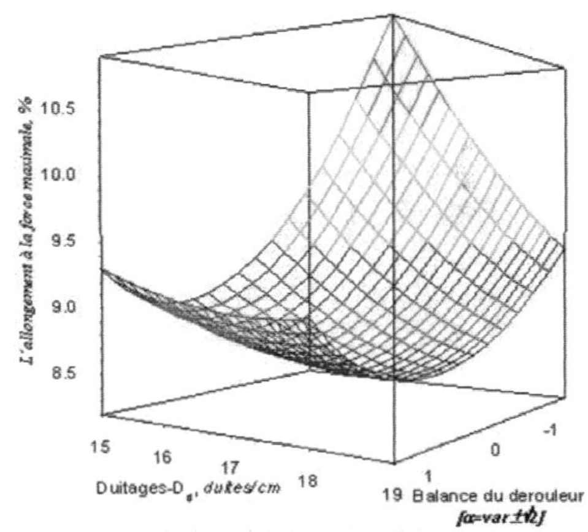

Fig.6, Lallongement par trame

The results show that at each stage of the weaving cycle the position of the roll corresponds to the instantaneous stress of the warp. There is a clear difference between the maximum level of the roll, which corresponds to the compaction of the weft, as well as the sinusoidal movement of the guide lever as a result of the periodic elongation of the main threads.
In the second stage, we studied the dynamic equilibrium of the controller at normal speed. Using an additional scale mounted on the vat, we marked the minimum and maximum levels of the roll for a short period of time. The changes in the machine settings follow the combinations of a planned factorial experiment with two independent factors: the torque and the lever ratio of the transmission element. The obtained results were used for the construction of the three-dimensional graph with two surfaces, fig. 9.

## 4. Conclusion

The interaction between warp and weft during the crisscrossing is not subject to direct observation. The planned experiment substitutes the impossible measurement of the physical parameters and indirectly describes the mechanical events in the fabric structure.

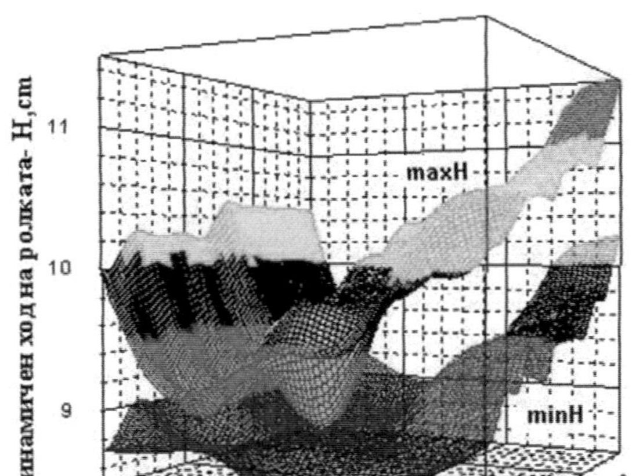


*Figure 9, Dynamic oscillations of the let-off motion roller*

The explored properties react differently to changes in settings. The stress state of the chain incorporates the machine impact. It was the only technological factor that directed certain inferences, valid in the field of experimentation of interest. The tension of the warp threads increases the maximum tensile force of the fabric, as well as the elongation per weft, but reduces the elongation per warp. This sequence of procedures is applicable to the optimization of fabric articles.
The surfaces show the averages of the two end levels of the guide lever for a relatively large number of consecutive weaving cycles. The reactions of the mechanism of the main regulator clearly differ depending on the change of the machine settings.

Discussion: main feature studied is the amplitude of the roll. From the position of the roller in static conditions, the instantaneous influence of the respective mechanism can be estimated.
At normal speed, the effects of the kinetic moments of the inertial elements are manifested. The amplitude of the dynamic oscillations of the roller is represented by the difference between the two surfaces. The change of factors leads to the simultaneous change of the amplitude and the level of dynamic equilibrium. The increase in torque leads to a decrease in the level of the roller, as well as a reduced cyclic movement of the guide lever. The reduced amplitude is explained by the increased moment of inertia of the lever and the increased internal deformation and reduced viscoelastic properties of the rheological element.
Increasing the lever ratio mainly affects the level of dynamic equilibrium and in this case, the level of the roller increases under the predominant action of the base voltage.

**Key words:** weaving mechanics, let-off motion, rheological model

**References:**

[1] Lamoitier Paul, Traité théorique et pratique de tissage, Librairie polytechnique Ch. Béranger, Paris et Liège, 1923.
[2] Droesbeke J-J., Fine J., Saporta G., Plans d'expériences (Applications à l'entreprise), ISBN 2-7108-0733-5, Editions Technip Paris, 1997.
[3] Kovačevič Stana, and co, Influence of Warp Loading on Weaving Machines upon Yarn Deformation, TRJ, v.70(7), p.603-610 (2000)
[4] Beitelschmidt M., Simulation of warp and cloth forces in weaving machines, Melliand International, v.6, issue 1, p.43-46, march 2000
[5] MAV, Notice d'entretien et de réglage de la machine à tisser « MAV-STANDARD » pour le type des machines : MAV-S-RPC-185, (Service Tissage Après -Vente, SACM, 1, rue de la Fonderie, 68054 Mulhouse CEDEX, France, 1977).
[6] Debaen Jacques, Mécanique et chaleur, B5, Techniques de l'ingénieur, ISSN 0399-4104, Paris, 1983

# AUTHOR INDEX